STUFF MATTERS

STUFF MATTERS

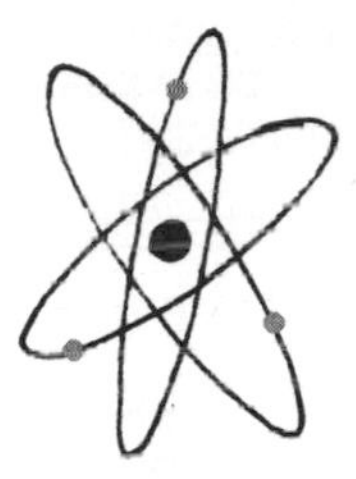

EXPLORING *the* MARVELOUS MATERIALS *that* SHAPE OUR MAN-MADE WORLD

MARK MIODOWNIK

HOUGHTON MIFFLIN HARCOURT

BOSTON NEW YORK 2014

First U.S. edition 2014

First published in the United Kingdom by Penguin Books Ltd 2013

www.hmhco.com

Library of Congress Cataloging-in-Publication Data
Miodownik, Mark, author.
Stuff matters : exploring the marvelous materials that shape our man-made world / Mark Miodownik. — First U.S. edition.
pages cm
Reprint of: London : Penguin, 2013.
ISBN 978-0-544-23604-2 (hardback)
1. Materials science — Popular works. I. Title.
TA403.2.M56 2014
620.1'1 — dc23
2013047575

Book design by Patrick Barry

Printed in the United States of America
DOC 10 9 8 7 6 5 4 3 2 1

For Ruby, Lazlo, and Ida

CONTENTS

INTRODUCTION

AS I STOOD ON a train bleeding from what would later be classified as a thirteen-centimeter stab wound, I wondered what to do. It was May 1985, and I had just jumped on to a London Tube train as the door closed, shutting out my attacker, but not before he had slashed my back. The wound stung like a very bad paper cut, and I had no idea how serious it was, but being a schoolboy at the time, embarrassment overcame any sort of common sense. So instead of getting help, I decided the best thing would be to sit down and go home, and so, bizarrely, that is what I did.

To distract myself from the pain, and the uneasy feeling of blood trickling down my back, I tried to work out what had just happened. My assailant had approached me on the platform asking me for money. When I shook my head he got uncomfortably close, looked at me intently, and told me he had a knife. A few specks of spit from his mouth landed on my glasses as he said this. I followed his gaze down to the pocket of his blue anorak. I had a gut feeling that it was just his index finger that was creating the pointed bulge. Even if he did have a knife, it must be so small to fit in that pocket that there was no way it could do me much damage. I owned penknives myself and knew that such a knife would find it very hard to pierce the several layers that I was wearing: my leather jacket, of which I was very proud, my gray wool school blazer beneath it, my nylon V-neck sweater, my cotton white shirt with obligatory striped school tie half knotted, and cotton vest. A plan formed quickly in my head: keep him talking and then push past him on to the train as the doors were closing. I could see the train arriving and was sure he wouldn't have time to react.

Funnily enough I was right about one thing: he didn't have a knife. His weapon was a razor blade wrapped in tape. This tiny piece of steel, not much bigger than a postage stamp, had cut through five layers of my clothes, and then through the epidermis and dermis of my skin in one slash without any problem at all. When I saw that weapon in the police station later, I was mesmerized. I had seen razors before of course, but now I realized that I didn't know them at all. I had just started shaving at the time, and had only seen them encased in friendly orange plastic in the form of a Bic safety razor. As the police quizzed me about the weapon, the table between us wobbled and the razor blade sitting on it wobbled too. In doing so it glinted in the fluorescent lights, and I saw clearly that its steel edge was still perfect, unaffected by its afternoon's work.

Later I remember having to fill in a form, with my parents anxiously sitting next to me and wondering why I was hesitating. Perhaps I had forgotten my name and address? In truth I had started to fixate on the staple at the top of the first page. I was pretty sure this was made of steel too. This seemingly mundane piece of silvery metal had neatly and precisely punched its way through the paper. I examined the back of the staple. Its two ends were folded snugly against one another, holding the sheaf of papers together in a tight embrace. A jeweler could not have made a better job of it. (Later I found out that the first stapler was hand-made for King Louis XV of France with each staple inscribed with his insignia. Who would have thought that staplers have royal blood?) I declared it "exquisite" and pointed it out to my parents, who looked at each other in a worried way, thinking no doubt that I was having a nervous breakdown.

Which I suppose I was. Certainly something very odd was going on. It was the birth of my obsession with materials—starting with steel. I suddenly became ultra-sensitive to its being present everywhere. I saw it in the tip of the ballpoint pen I was using to fill out the police form; it jangled at me from my dad's key ring while he waited, fidgeting; later that day it sheltered and took me home, covering the outside of our car in a layer no thicker than a postcard. Strangely, I felt that our steel Mini, usually so noisy, was on its best behavior that day, materially apologizing for the stabbing incident. When we got home I sat down next to my dad at the kitchen table, and we ate my mum's soup together in silence. Then I paused, realizing I even had a piece of steel in my mouth. I consciously sucked the stainless steel spoon I had been eating my soup with, then took it out and studied its bright shiny appearance, so shiny that I could even see a distorted reflection of myself in it. "What is this stuff?" I said, waving the spoon at my dad. "And why doesn't it taste of anything?" I put it back in my mouth to check, and sucked it assiduously.

Then a million questions poured out. How is it that this one material does so much for us, and yet we hardly talk about it? It is an intimate character in our lives—we put it in our mouths, use it to get rid of unwanted hair, drive around in it—it is our most faithful friend, and yet we hardly know what makes it tick. Why does a razor blade cut while a paper clip bends? Why are metals shiny? Why, for that matter, is glass transparent? Why does everyone seem to hate concrete but love diamond? And why is it that chocolate tastes so good? Why does any material look and behave the way it does?

Since the stabbing incident, I have spent the vast majority of my time obsessing about materials. I've studied materials science at Oxford University, I've earned a PhD in jet engine alloys, and I've worked as a materials scientist and engineer in some of the most advanced laboratories around the world. Along the way, my fascination with materials has continued to grow—and with it my collection of extraordinary samples of them. These samples have now been incorporated into a vast library of materials built together with my friends and colleagues Zoe Laughlin and Martin Conreen. Some are impossibly exotic, such as a piece of NASA aerogel, which being 99.8 percent air resembles solid smoke; some are radioactive, such as the uranium glass I found at the back of an antique shop in Australia; some are small but stupidly heavy, such as ingots of the metal tungsten extracted painstakingly from the mineral wolframite; some are utterly familiar but have a hidden secret, such as a sample of self-healing concrete. Taken together, this library of more than a thousand materials represents the ingredients that built our world, from our homes, to our clothes, to our machines, to our art. The library is now located and maintained at the Institute of Making which is part of University College London. You could rebuild our civilization from the contents of this library, and destroy it too.

Yet there is a much bigger library of materials containing millions of materials, the biggest ever known, and it is growing at an exponential rate: the man-made world itself. Consider the photograph on page xiv. It pictures me drinking tea on the roof of my flat. It is unremarkable in most ways, except that when you look carefully it provides a catalog of the stuff from which our whole civilization is made. This stuff is important. Take away the concrete, the glass, the textiles, the metal, and the other materials from the scene, and I am left naked, shivering in midair. We may like to think of ourselves as civilized, but that civilization is in large part bestowed by material wealth. Without this stuff, we would quickly be confronted by the same basic struggle for survival that animals are faced with. To some extent, then, what allows us to behave as humans are our clothes, our homes, our cities, our stuff, which we animate through our customs and language. (This becomes clear if you ever visit a disaster zone.) The material world is not just a display of our technology and culture, it is part of us. We invented it, we made it, and in turn it makes us who we are.

The fundamental importance of materials to us is apparent from the names we have used to categorize the stages of civilization—the Stone Age, Bronze Age, and Iron Age—with each new era of human existence being brought about by a new material. Steel was the defining material of the Victorian era, allowing engineers to give full rein to their dreams of creating suspension bridges, railways, steam engines, and passenger liners. The great engineer Isambard Kingdom Brunel used it to transform the landscape and sowed the seeds of modernism. The twentieth century is often hailed as the Age of Silicon, after the breakthrough in materials science that ushered in the silicon chip and the information revolution. Yet this is to overlook the kaleidoscope of other new materials that also revolutionized modern living at that time. Architects took mass-produced sheet glass and combined it with structural steel to produce skyscrapers that invented a new type of city life. Product and fashion designers adopted plastics and transformed our homes and dress. Polymers were used to produce celluloid and ushered in the biggest change in visual culture for a thousand years: the cinema. The development of aluminum alloys and nickel superalloys enabled us to build jet engines and fly cheaply, thus accelerating the collision of cultures. Medical and dental ceramics allowed us to rebuild ourselves and redefine disability and aging—and, as the term *plastic surgery* implies, materials are often the key to new treatments used to repair our faculties (hip replacements) or enhance our features (silicone implants for breast enlargement). Gunther von Hagens's *Body Worlds* exhibitions also testify to the cultural influence of new biomaterials, inviting us to contemplate our physicality in both life and death.

This book is for those who want to decipher the material world we have constructed and find out where these materials came from, how they work, and what they say about us. The materials themselves are often surprisingly obscure, despite being all around us. On first inspection they rarely reveal their distinguishing features and often blend into the background of our lives. Most metals are shiny and gray; how many people can spot the difference between aluminum and steel? Woods are clearly different from

each other, but how many people can say why? Plastics are confusing; who knows the difference between polythene and polypropylene? I have chosen as my starting point and inspiration for the contents of this book the photo of me on my roof. I have picked ten materials found in that photo to tell the story of stuff. For each I try to uncover the desire that brought it into being, I decode the materials science behind it, I marvel at our technological prowess in being able to make it, but most of all I try to express why it matters.

Along the way, we find that, as with people, the real differences between materials are deep below the surface, a world that is shut off from most unless they have access to sophisticated scientific equipment. So to understand materiality, we must necessarily journey away from the human scale of experience into the inner space of materials. It is at this microscopic scale that we discover why some materials smell and others are odorless; why some materials can last for a thousand years and others become yellow and crumble in the sun; how it is that some glass can be bulletproof, while a wine glass shatters at the slightest provocation. The journey into this microscopic world reveals the science behind our food, our clothes, our gadgets, our jewelry, and of course our bodies.

But while the physical scale of this world is much smaller, we will find that its timescale is often dramatically bigger. Take, for example, a piece of thread, which exists at the same scale as hair. Thread is a man-made structure at the limit of our eyesight that has allowed us to make ropes, blankets, carpets, and, most importantly, clothes. Textiles are one of the earliest man-made materials. When we wear a pair of jeans, or any other piece of clothing, we are wearing a miniature woven structure, the design of which is much older than Stonehenge. Clothes have kept us warm and protected for all of recorded history, as well as keeping us fashionable. But they are high-tech too. In the twentieth century we learned how to make space suits from textiles strong enough to protect astronauts on the Moon; we made solid textiles for artificial limbs; and from a personal perspective, I am happy to note the development of stab-proof underwear woven from a synthetic high-strength fiber called Kevlar. This evolution of our materials technologies over thousands of years is something I return to again and again in this book.

Each new chapter presents not just a different material but a different way of looking at it—some take a primarily historical perspective, others a more personal one; some are conspicuously dramatic, others more coolly scientific; some emphasize a material's cultural life, others its astonishing technical abilities. All the chapters are a unique blend of these approaches, for the simple reason that materials and our relationships with them are too diverse for a single approach to suit them all. The field of materials science provides the most powerful and coherent framework for understanding them technically, but there is more to materials than the science. After all, everything is made from something, and those who make things—artists, designers, cooks, engineers, furniture makers, jewelers, surgeons, and so on—all have a different understanding of the practical, emotional, and sensual aspect of their materials. It is this diversity of material knowledge that I have tried to capture.

For instance, the chapter on paper is in the form of a series of snapshots not just because paper comes in many forms but because it is used by pretty much everyone in a myriad of different ways. The chapter on biomaterials, on the other hand, is a journey deep into the interstices of our material selves: our bodies, in fact. This is a terrain that is rapidly becoming the Wild West of materials science, where new materials are opening up a whole new area of bionics, allowing the body to be rebuilt with the help of bio-implants designed to mesh "intelligently" with our flesh and blood. Such materials have profound ramifications for society as they promise to change fundamentally our relationship with ourselves.

Because everything is ultimately built from atoms, we cannot avoid talking about the rules that govern them, which are described by the theory known as quantum mechanics. This means that, as we enter the atomic world of the small, we must abandon common sense utterly, and talk instead of wave functions and electron states. A growing number of materials are being designed from scratch at this scale, and can perform seemingly impossible tasks. Silicon chips designed using quantum mechanics have already brought about the information age. Solar cells designed in a similar way have the potential to solve our energy problems using only sunshine. But we are not there yet, and still rely on oil and coal. Why? In this book I try to shed some light on the limits of what we can hope to achieve by examining the great new hope in this arena: graphene.

The central idea behind materials science is that changes at these invisibly small scales impact a material's behavior at the human scale. It is this process that our ancestors stumbled upon to make new materials such as bronze and steel, even though they did not have the microscopes to see what they were doing—an amazing achievement. When you hit a piece of metal you are not just changing its shape, you are changing the inner structure of the metal. If you hit it in a particular way, this inner structure changes in such a way that the metal gets harder. Our ancestors knew this from experience even though they didn't know why. This gradual accumulation of knowledge got us from the Stone Age to the twentieth century before any real appreciation of the structure of materials was understood. The importance of that empirical understanding of materials, encapsulated in such crafts as the blacksmith's, remains: we know almost all of the materials in this book with our hands as well as our heads.

This sensual and personal relationship with stuff has fascinating consequences. We love some materials despite their flaws, and loathe others even if they are more practical. Take ceramic. It is the material of dining: of our plates, bowls, and cups. No home or restaurant is complete without this material. We have been using it since we invented agriculture thousands of years ago, and yet ceramics are chronically prone to chip, crack, and shatter at the most inconvenient times. Why haven't we moved to tougher materials, such as plastic or metal for our plates and cups? Why have we stuck with ceramic despite its mechanical shortcomings? This type of question is studied by a vast variety of academics, including archaeologists and anthropologists, as well as designers and artists. But there is also a scientific discipline especially dedicated to systematically investigating our sensual interactions with materials. This discipline, called psychophysics, has made some very interesting discoveries. For instance, studies of "crispness" have shown that the sound created by certain foods is as important to our enjoyment of them as their taste. This has inspired some chefs to create dishes with added sound effects. Some potato chip manufacturers, meanwhile, have increased not just the crunchiness of their chips but the noisiness of the chip bag itself. I explore the psychophysical aspects of materials in a chapter on chocolate and show that it has been a major driver of innovation for centuries.

This book is by no means an exhaustive survey of materials and their relationship to human culture, but rather a snapshot of how they affect our lives, and how even the most innocuous of activities like drinking tea on a roof is founded on a deep material complexity. You don't have to go into a museum to wonder at how history and technology have affected human culture; their effects are all around you now. Most of the time we ignore them. We have to: we would be treated as lunatics if we spent the whole time running our fingers down a concrete wall and sighing. But there are times for such contemplation: being stabbed in a Tube station was one of them for me, and I hope this book provides another for you.

STUFF MATTERS

1 INDOMITABLE

I HAD NEVER BEEN asked to sign a non-disclosure agreement in the bathroom of a pub before, so it came as something of a relief to discover that this was all that Brian was asking me to do. I had met Brian for the first time only an hour earlier. We were in Sheehan's, a pub in Dun Laoghaire that wasn't far from where I worked at the time in Dublin. Brian was a red-faced man in his sixties with a walking stick for his bad leg. He was smartly dressed in a suit and had thinning gray hair with a yellowish tinge. He chain-smoked Silk Cut cigarettes. Once Brian found out that I was a scientist he guessed rightly that I would be interested to hear stories of his life in London in the 1970s, when he was in the right place at the right time to trade Intel 4004 silicon chips, which he imported in boxes of twelve thousand for £1 each and sold in small batches to the fledgling computer industry for £10 each. When I mentioned that I was researching metal alloys in the Mechanical Engineering Department of University College Dublin, he looked pensive and was quiet for the first time. I took this as an opportune moment to head to the bathroom.

The non-disclosure agreement was scrawled on a piece of paper which he had clearly just ripped out of his notebook. The contents were brief. They stated that he was going to explain his invention to me but I had to keep it confidential. In return he was to pay me one Irish pound. I asked him to tell me more, but he comically mimed the zipping of his lips. I wasn't quite sure why we had to have this conversation in a bathroom stall. Over his shoulder I saw other drinkers come in and out of the bathroom. I wondered if I should cry out for help. Brian searched in his jacket and got out a pen. A scruffy pound note emerged from his jeans. He was very insistent.

I signed the paper against the graffiti-daubed wall. He signed too, gave me the pound, and the slip of paper became a legal document.

Back by the bar with our drinks, I listened as Brian explained that he had invented an electronic machine that sharpened blunt razor blades. This would revolutionize the shaving business, he explained, because people would need to own only one razor in their lives. At a stroke it would put the billion-dollar industry out of business, make him an exceptionally rich man, and reduce consumption of Earth's mineral wealth. "How about that?" he said, taking a triumphant gulp of his pint.

I eyed him with suspicion. Sooner or later every scientist has his ear bent by someone with a crackpot idea for an invention. In addition, razor blades were a sensitive subject for me. I felt prickly and uncomfortable as I became aware of the long scar down my back, the result of my encounter on the platform at Hammersmith station. But I gestured for him to continue and kept listening . . .

It is an odd fact that steel was not understood by science until the twentieth century. Before that, for thousands of years, the making of steel was handed down through the generations as a craft. Even in the nineteenth century, when we had an impressive theoretical understanding of astronomy, physics, and chemistry, the making of iron and steel on which our Industrial Revolution was based was achieved empirically—through intuitive guesswork, careful observation, and a huge slice of luck. (Could Brian have had such a slice of luck and simply stumbled upon a revolutionary new process for sharpening razor blades? I found that I wasn't prepared to dismiss the idea.)

During the Stone Age, metal was extremely rare and highly prized, since the only sources of it on the planet were copper and gold, which occur naturally, if infrequently, in the Earth's crust (unlike most metals, which have to be extracted from ores). Some iron existed too, most of it having fallen from the sky in the form of meteorites.

Radivoke Lajic, who lives in northern Bosnia, is a man who knows all about strange bits of metal falling from the sky. Between 2007 and 2008 his house was hit by no fewer than five meteorites, which is statistically so hugely unlikely that his claim that aliens were targeting him seems almost reasonable. Since Lajic went public with his suspicions in 2008, his house has been hit by another meteorite. The scientists investigating the strikes have confirmed that the rocks hitting his house are real meteorites and are studying the magnetic fields around his house to try to explain the extremely unusual frequency of them.

Radivoke Lajic and the five meteorites that have hit his house since 2007.

In the absence of copper, gold, and meteoric iron, our ancestors' tools during the Stone Age were made of flint, wood, and bone. Anyone who has ever tried to make anything with these kinds of tools knows how limiting they are: if you hit a piece of wood it either splinters, cracks, or snaps. The same is true of rock or bone. Metals are fundamentally different from these other materials because they can be hammered into shape: they flow, they are malleable. Not only that, they get stronger when you hit them; you can harden a blade just by hammering it. And you can reverse the process simply by putting metal in a fire and heating it up, which will cause it to get softer. The first people to discover these properties ten thousand years ago had found a material that was almost as hard as a rock but behaved like a plastic and was almost infinitely reusable. In other words, they had discovered the perfect material for tools, and in particular cutting tools like axes, chisels, and razors.

This ability of metals to transform from a soft to a hard material must have seemed like magic to our ancient ancestors. It was magic to Brian too, as I soon found out. He explained that he had invented his machine by trial and error, with no real appreciation of the physics and chemistry at play, and yet it seemed that he had somehow succeeded. What he wanted from me was to measure the sharpness of the razors before and after they had been through his process. Only this evidence would allow him to begin serious business discussions with the razor companies.

I explained to Brian that it would take more than a few meas-

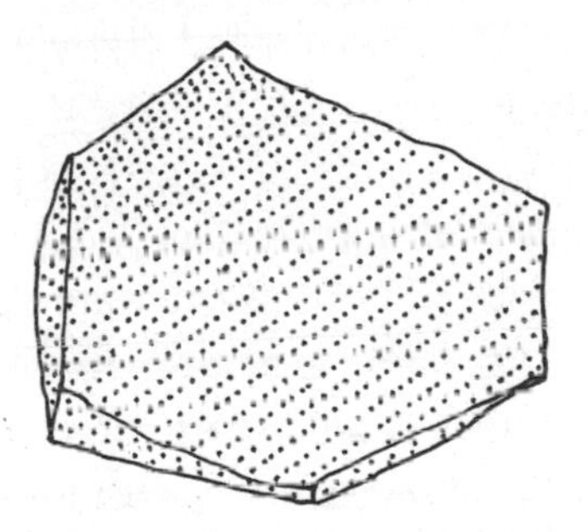

A metal crystal, such as exists inside a razor. The rows of dots represent atoms.

urements for them to take him seriously. The reason is that metals are made from crystals. The average razor blade contains billions of them, and in each of these crystals the atoms are arranged in a very particular way, a near-perfect three-dimensional pattern. The bonds between the atoms hold them in place and also give the crystals their strength. A razor gets blunt because the many collisions with hairs that it encounters force bits of these crystals to rearrange themselves into a different shape, making and breaking bonds and creating tiny dents in the smooth razor edge. Resharpening a razor through some electronic mechanism, as he proposed, would have to reverse this process. In other words, it would have to move atoms around to rebuild the structure that had been destroyed. To be taken seriously, Brian would need not just evidence of such rebuilding at the scale of the crystals but a plausible explanation at the atomic scale of the mechanism by which it worked. Heat, whether electrically produced or not, usually has a different effect than the one he was claiming: it softens metal crystals, I explained. Brian was adamant that his electronic machine wasn't heating the steel razors.

It may be odd to think that metals are made of crystals, because our typical image of a crystal is of a transparent and highly faceted gemstone such as a diamond or emerald. The crystalline nature of metals is hidden from us because metal crystals are opaque, and in most cases microscopically small. Viewed through an electron microscope, the crystals in a piece of metal look like crazy paving, and inside those crystals are squiggly lines — these are dislocations. They are defects in the metal crystals, and represent deviations in the otherwise perfect crystalline arrangement of the atoms — they are atomic disruptions that shouldn't be there. They sound bad, but they turn out to be very useful. Dislocations are what make metals so special as materials for tools, cutting edges, and ultimately the razor blade, because they allow the metal crystals to change shape.

You don't need to use a hammer to experience the power of dislocations. When you bend a paper clip, it is in fact the metal crystals that are bending. If they didn't bend, the paper clip would be brittle and snap like a stick. This plastic behavior is achieved by the dislocations moving within the crystal. As they move they transfer small bits of the material from one side of the crystal to the other. They do this at the speed of sound. As you bend a paper

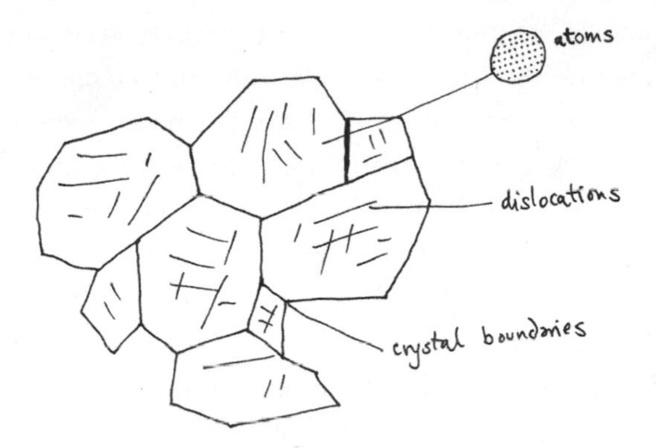

I have only shown a few dislocations in this sketch to make them easy to see. Normal metals have enormous numbers of dislocations which overlap and intersect.

clip, you are causing approximately 100,000,000,000,000 dislocations to move at a speed of thousands of hundreds of meters per second. Although each one only moves a tiny piece of the crystal (one atomic plane in fact), there are enough of them to allow the crystals to behave like a super-strong plastic rather than a brittle rock.

The melting point of a metal is an indicator of how tightly the metal atoms are stuck together and so also affects how easily the dislocations move. Lead has a low melting point and so dislocations move with consummate ease, making it a very soft metal. Copper has a higher melting point and is stronger. Heating metals allows dislocations to move about and reorganize themselves, with one of the outcomes being that it makes metals softer.

Discovering metals was an important moment in pre-history, but it didn't solve the fundamental problem that there wasn't very much metal around. One option, clearly, was to wait for some more to drop from the sky, but this requires a huge amount of patience (a few kilograms fall to the surface of the Earth every year, but mostly into the oceans). At some point humans made the discovery that would end the Stone Age and open the door to a seemingly unlimited supply of the stuff. They discovered that a certain greenish rock, when put into a very hot fire and surrounded by red-hot embers, turns into a shiny piece of metal. This greenish rock was malachite, and the metal was, of course, copper. It must have been the most dazzling revelation. Suddenly the discoverers were surrounded not by dead inert rock but by mysterious stuff that had an inner life.

They would have been capable of performing this transformation with only a few particular types of rock, such as malachite, because getting it to work reliably depends not just on identifying these rocks but also on carefully controlling the chemical conditions of the fire. But they must have suspected that those rocks that didn't work, that remained obstinately rock-like however hot the fire became, had hidden secrets. They were right. It's a process that works for many minerals, although it would be thousands of years before an understanding of the chemistry required (controlling the chemical reactions between the rock and the gases created in the fire) led to the next real breakthrough in smelting.

In the meantime, from around 5000 BC, early metalsmiths used trial and error to hone the process of the production of copper. The making of copper tools initiated a spectacular growth in human technology, being instrumental in the birth of other technologies, cities, and the first great civilizations. The pyramids of Egypt are an example of what became possible once there were plentiful copper tools. Each block of stone in each pyramid was extracted from a mine and individually hand-carved using copper chisels. It is estimated that ten thousand tons of copper ore were mined throughout ancient Egypt to create the three hundred thousand chisels needed. It was an enormous achievement, without which the pyramids could not have been built, however many slaves were used, since it is not practical to carve rock without metal tools. It is all the more impressive given that copper is not the ideal material for cutting rock since it is not very hard. Sculpting a piece of limestone with a copper chisel quickly blunts the chisel. It is estimated that the copper chisels would have needed to be sharpened every few hammer blows in order for them to be useful. Copper is not ideal for razor blades for the same reason.

Gold is another relatively soft metal, so much so that rings are very rarely made from pure gold metal because they quickly scratch. But if you alloy gold, by adding a small percentage of other metals such as silver or copper, you not only change the color of the gold—silver making the gold whiter, and copper making the gold redder—you make the gold harder, much harder. This changing of the properties of metals by very small additions of other ingredients is what makes the study of metals so fascinating. In the case of gold alloys, you might wonder where the silver atoms go. The answer is that they sit inside the gold crystal structure, taking the place of a gold atom, and it is this atom substitution inside the crystal lattice of the gold that makes it stronger.

Alloys tend to be stronger than pure metals for one very simple reason: the alloy atoms have a different size and chemistry from the host metal's atoms, so when they sit inside the host crystal they cause all sorts of mechanical and electrical disturbances that add up to one crucial thing: they make it more difficult for dislocations to move. And if dislocations find it difficult to move, then the metal is stronger, since it's harder for the metal crystals to change shape. Alloy design is thus the art of preventing the movement of dislocations.

These atom substitutions happen naturally inside other crystals too. A crystal of aluminum oxide is colorless if pure but becomes blue when it contains impurities of iron atoms: it is the gemstone called sapphire. Exactly the same aluminum oxide crystal containing impurities of chromium is the gem called ruby.

The ages of civilization, from the Copper Age to the Bronze Age

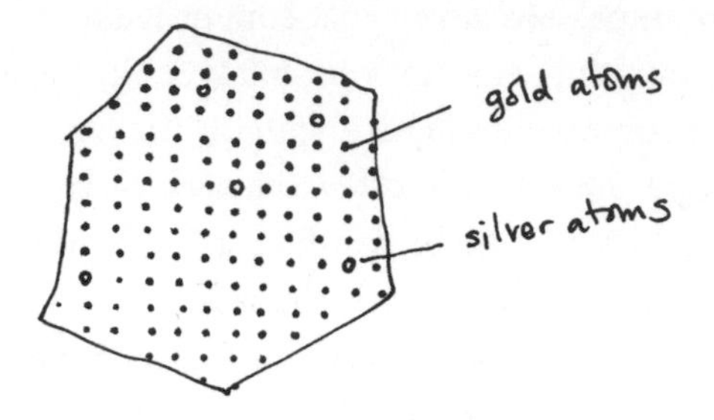

Gold alloyed with silver at the atomic scale, showing how the silver atoms replace some of the gold atoms in the crystal.

to the Iron Age, represent a succession of stronger and stronger alloys. Copper is a weak metal, but naturally occurring and easy to smelt. Bronze is an alloy of copper, containing small amounts of tin or sometimes arsenic, and is much stronger than copper. So, if you had copper and you knew what you were doing, for very little extra effort you could create weapons and razors ten times stronger and harder than copper. The only problem is that tin and arsenic are extremely rare. Elaborate trade routes evolved in the Bronze Age to bring tin from places such as Cornwall and Afghanistan to the centers of civilization in the Middle East for precisely this reason.

Modern razors are also made from an alloy but, as I explained to Brian, it is a very special sort of alloy, the existence of which puzzled our ancestors for thousands of years. Steel, the alloy of iron and carbon, is even stronger than bronze, with ingredients that are much more plentiful: pretty much every bit of rock has some iron in it, and carbon is present in the fuel of any fire. Our ancestors didn't realize that steel was an alloy—that carbon, in the form of charcoal, was not just a fuel to be used for heating and reshaping iron but could also get inside the iron crystals in the process. Carbon doesn't do this to copper during smelting, nor to tin or bronze, but it does to iron. It must have been incredibly mysterious—and only now with a knowledge of quantum mechanics can we truly explain why it happens (the carbon in steel doesn't take the place of an iron atom in the crystal, but is able to squeeze in between the iron atoms, creating a stretched crystal).

There is another problem, too. If iron becomes alloyed with too much carbon—if, for instance, it contains 4 percent carbon instead of 1 percent carbon—then it becomes extremely brittle and essentially useless for tools and weapons. This is a major obstacle because inside a fire there is rather a lot of carbon around. Leave the iron in too long, or allow it to become liquid in the fire, and a huge amount of carbon enters the metal crystals, making the alloy very brittle. Swords made from this high-carbon steel snap in battle.

Until the twentieth century, when the alloying process was first fully explained, no one understood why some steelmaking processes worked and others didn't. They were established by trial and error, and those that were successful were handed down to the next generation and were often trade secrets. But even if they were stolen, they were so complicated that the chances of successfully reproducing someone else's steelmaking process were very low. Certain metallurgical traditions in certain cultures became known for making extremely high-quality steel, and such civilizations thrived.

In 1961 Professor Richmond from Oxford University discovered a pit that had been dug by the Romans in AD 89. It contained 763,840 small two-inch nails, 85,128 medium nails, 25,088 large nails, and 1,344 extra-large sixteen-inch nails. The hoard was of iron and steel and not gold, which most people would have found bitterly disappointing. But not Professor Richmond. Why, he asked himself, would a Roman legion bury seven tons of iron and steel?

The legion had been occupying the advance headquarters of Agricola in a place called Inchtuthil in Scotland. This was at the outer reaches of the Roman Empire, and their mission was to protect its border from what they saw as the savage tribes who threatened it: the Celts. The legion of five thousand men occupied the region for six years before retreating and, in the process, abandoning their fort. They made great efforts to leave behind nothing that could help their enemies. They smashed all food and drink containers and burned the fort to the ground. But they weren't satisfied with this. In the ashes were the steel nails that had held the fort together, and they were far too valuable to be left to the tribes that had driven them out. Iron and steel were the materials that enabled the Romans to build aqueducts, ships, and swords; they allowed them to engineer an empire. Leaving the nails to their enemies would have been as useful as leaving a cache of weapons, so they buried them in a pit before marching south. As well as their weapons and armor, among the few, smaller steel objects that they probably did take with them were *novacili,* an object that epitomized their civilized approach to life: the Roman razor blade. These *novacili,* and the barbers who wielded them, allowed the Romans to retreat clean-shaven, groomed in order to distinguish themselves from the savage hordes that had driven them out.

The mystique that surrounded steelmaking engendered various myths, and the unification and restoration of order to Britain in the wake of the Roman retreat was symbolized by one of the most enduring of these: Excalibur, the legendary sword of King Arthur, sometimes attributed with magical powers and associated with the rightful sovereignty of Britain. At a time when swords regularly snapped in battle, leaving a knight defenseless, it is easy to see why a high-quality steel sword wielded by a strong warrior came to represent the rule of civilization over chaos. The fact that the process of making steel was, necessarily, highly ritualized also helps to explain why this material came to be associated with magic.

This was nowhere more true than in Japan, where the forging of a samurai blade took weeks and was part of a religious ceremony. The *Ama-no-Murakumo-no-Tsurugi* ("Sword of the Gathering Clouds of Heaven") is a legendary Japanese sword which allowed the great warrior Yamato Takeru to control the wind and defeat all his enemies. Despite the fantastic stories and rituals, the idea that some swords could be made ten times stronger and sharper than other swords was not just a myth, but a reality. By the fifteenth century AD the sword steel made by the samurai of Japan was the best the world had ever seen and remained preeminent for five hundred years until the advent of metallurgy as a science in the twentieth century.

These samurai swords were made from a special type of steel called *tamahagane,* which translates as "jewel steel," made from the volcanic black sand of the Pacific (this consists mostly of an iron ore called magnetite, the original material for the needle of compasses). This steel is made in a huge clay vessel four feet tall, four feet wide, and twelve feet long called a *tatara.* The vessel is "fired"—hardened from molded clay into a ceramic—by lighting a fire inside it. Once fired, it is packed meticulously with layers of black sand and black charcoal, which are consumed in the ceramic furnace. The process takes about a week and requires constant attention from a team of four or five people, who make sure that the temperature of the fire is kept high enough by pumping air into the *tatara* using a manual bellows. At the end the *tatara* is broken open and the *tamahagane* steel is dug out of the ash and remnants of sand and charcoal. These lumps of discolored steel are very unprepossessing, but they have a whole range of carbon content, some of it very low and some of it high.

The samurai innovation was to be able to distinguish high-carbon steel, which is hard but brittle, from low-carbon steel, which is tough but relatively soft. They did this purely by how it looked, how it felt in their hands, and how it sounded when struck. By separating the different types of steel, they could make sure that the low-carbon steel was used to make the center of the sword. This gave the sword an enormous toughness, almost a chewiness, meaning that the blades were unlikely to snap in combat. On the edge of the blades they welded the high-carbon steel, which was brittle but extremely hard and could therefore be made very sharp.

By using the sharp high-carbon steel as a wrapper on top of the tough low-carbon steel they achieved what many thought impossible: a sword that could survive impact with other swords and armor while remaining sharp enough to slice a man's head off. The best of both worlds.

No one could create stronger and harder steel than the samurai until the Industrial Revolution. When at this time European countries first started to build structures on a larger, more ambitious scale—such as railways, bridges, and ships—they used cast iron, because it could be made in large quantities and poured into molds. Unfortunately it was extremely prone to fracture under certain conditions. As engineering became more ambitious, those conditions came about more often.

One of the worst accidents occurred in Scotland. On the night of December 28, 1879, the world's longest bridge, the cast-iron Tay Rail Bridge, collapsed during a powerful winter gale. A train carrying seventy-five passengers plunged into the River Tay, killing all of them. The disaster confirmed what many suspected, that iron just wasn't up to the job. What was needed was the ability not just to make steel as strong as samurai swords but to mass-produce it.

One day a Sheffield-based engineer named Henry Bessemer stood up at a meeting of the British Association for the Advancement of Science and announced he had done it. His process didn't require the elaborate procedures of the samurai. He could create tons of liquid steel. It was a revolution in the making.

The Bessemer process was ingeniously simple. It involved blowing air through the molten iron, so that the oxygen in the air would react with the carbon in the iron and remove it as carbon dioxide gas. It required a knowledge of chemistry that for the first time put steelmaking on a scientific footing. Moreover, the reaction between the oxygen and the carbon was extremely violent and gave off a lot of heat. This heat raised the temperature of the steel, keeping it hot and liquid. The process was straightforward and could be used on an industrial scale; it was the answer.

The only problem with the Bessemer process was that it didn't work. Or at least that was what everyone who tried it said. Soon, angry steelmakers, who had bought the license from Bessemer and invested large sums of money in equipment only to produce brittle iron, started asking for their money back. He had no answers for them. He didn't really understand why the process was successful sometimes and unsuccessful at others, but he continued to work on his technology, and with the help of the British metallurgist Robert Forester Mushet he adapted his technique. Rather than trying to remove the carbon until just the right amount was left, about 1 percent, Mushet suggested removing all the carbon and then adding 1 percent carbon back in. This worked and was repeatable.

Of course, when Bessemer tried to interest the world in this new process, the other steelmakers ignored him, assuming that it was yet another swindle. They insisted that it was impossible to create steel from liquid iron, and that Bessemer was a con artist. In the end he saw no option but to set up his own steel works and just start making the stuff himself. After a few years the firm of Henry Bessemer & Co. was manufacturing steel so much more cheaply and in such larger quantities than his rival firms that they were eventually forced to license his process, in the end making him extremely rich and ushering in the machine age.

Could Brian be another Bessemer? Could he have stumbled across a process for reorganizing the metal crystal structure at the tip of a razor blade through the action of electric or magnetic fields, a process he didn't understand but that worked nevertheless? There are many stories of those who have laughed at visionaries only to be embarrassed by their subsequent success. Many laughed at the idea that heavier-than-air flying machines were possible, and yet we all fly around in them. Likewise, television, mobile phones, computers—all have emerged from a cloud of derision.

Until the twentieth century, steel razors and surgical knives were extremely expensive. They had to be hand-made from the highest-grade steel since only this type of steel could be sharpened sufficiently to cut facial hair cleanly and effortlessly, without snagging. (Anyone who has used a blunt razor will know all too well how acutely painful even the slightest snag can be.) And because steel corrodes in the presence of air and water, cleaning the blades blunts them too, as the fine cutting edge literally rusts away. Thus for thousands of years the ritual of shaving began with the process of stropping: the act of sharpening the blade by playing it back and forth along a length of leather. You might think it not credible that a material as soft as leather can sharpen steel, and you would be right. It is the fine ceramic powder that is impregnated in the leather strop that does the sharpening. Traditionally a mineral called jewelers' rouge was used, but these days diamond powder is more common. The act of running the steel along the strop, in a flip-flop manner, causes the blade to meet the hard particles of diamond that are in the powder, which remove very small amounts of metal in the collision, restoring the delicately fine cutting edge.

But this changed when, in 1903, an American businessman called King Camp Gillette decided to use the new cheap industrial steel produced by the Bessemer process to create a disposable razor. This was to be the democratization of shaving. His vision was to eliminate the need to sharpen the blade by making it so cheap that when it became blunt you could simply throw it away. In 1903 Gillette sold 51 razors and 168 blades. The following year, he sold 90,884 razors and 123,648 blades. By 1915, the corporation had established manufacturing facilities in the United States, Canada, England, France, and Germany, and razor blade sales exceeded seventy million. The disposable steel razor became a permanent fixture of every bathroom, and people stopped needing to go to the barber's for a shave. And it has remained so: while there are any number of back-to-basics movements in food production, no one wants to have their hair cut with a copper knife or their face shaved with a blunt razor.

Gillette's business model was clever for many reasons, one of which was undoubtedly that even if the razors were not blunted through the act of shaving they would lose their edge quickly through rusting, assuring repeat business. But there was one further twist to the tale, an innovation so outrageously simple that it had to be discovered by accident.

In 1913, as the European powers were busily arming themselves for the First World War, Harry Brearley had the job of investigating metal alloys in order to create improved gun barrels. He was working in one of Sheffield, England's metallurgy labs, adding different alloying elements to steel, casting specimens, and then mechanically testing them for hardness. Brearley knew that steel was an alloy of iron and carbon, and he also knew that lots of other elements could be added to steel to improve or destroy its properties. No one at the time knew why, so he proceeded by trial and error, melting steels and adding different ingredients in order to discover their effects. One day it was aluminum, the next it was nickel.

Brearley made no progress. If a new specimen turned out not to be hard, he chucked it in the corner. His moment of genius came when after a month he walked through the lab and saw a bright glimmer in the pile of rusting specimens. Rather than ignoring it and going to the pub, he fished out this one specimen that had not rusted and realized its significance: he was holding the first piece of stainless steel the world had ever known.

Accidentally, by getting the ratios of two alloy ingredients right, carbon and chromium, he had managed to create a very special crystal structure in which the chromium and carbon atoms were both inserted inside the iron crystals. The addition of chromium had not made the steel harder, hence he had rejected the sample, but it had done something much more interesting. Normally when steel is exposed to air and water, the iron on the surface reacts to form iron(III) oxide, a red mineral commonly known as rust. When this rust flakes off, it exposes another layer of the steel to further corrosion, which is what makes rusting such a chronic problem for steel structures, hence the need to paint steel bridges and cars. But with chromium present something different happens. Like some hugely polite guest, it reacts with the oxygen before the host iron atoms can do so, creating chromium oxide. Chromium oxide is a transparent, hard mineral that sticks extremely well to steel. In other words, it doesn't flake off and you don't know it is there. Instead it creates an invisible, chemically protective layer over the whole surface of the steel. What's more, we now know that the protective layer is self-healing; when you scratch stainless steel, even though you break the protective barrier, it re-forms.

Brearley went on to try to make the world's first stainless steel knives, but immediately ran into problems. The resulting metal was not hard enough to make a sharp edge, and they were soon dubbed "knives that would not cut." This lack of hardness was, after all, the very reason that Brearley had rejected the alloy for use in gun barrels. Its lack of hardness allowed the alloy to do other things, though, which only became apparent later in the century—namely, it could be formed into complex shapes, leading eventually to one of the most influential pieces of sculpture, present in almost every house: the kitchen sink.

Stainless steel sinks are indomitable and gleaming and seem able to take anything that is thrown at them. In a world where we are keen to dispose of waste instantly and conveniently—from fat, to bleach, to acid—this material has really come through for us. It has ousted ceramic sinks from the kitchen, and would oust the ceramic bowl from the bathroom if we would let it, but we do not yet trust this new metal quite enough for that most intimate of waste disposal jobs.

Stainless steel is the very epitome of our modern age. It is clean-looking and shiny, appears almost indestructible, but is ultimately democratic: in less than a hundred years it has become the metal with which we are the most closely acquainted; after all, we put it in our mouths almost every day. For, in the end, Brearley did manage to create cutlery from stainless steel, and it's the transparent protective layer of chromium oxide that makes the spoon tasteless, since your tongue never actually touches the metal and your saliva cannot react with it; it has meant that we are one of the first generations who have not had to taste our cutlery. It is often used in architecture and art precisely because its bright surface appears uncorrodable. Anish Kapoor's *Cloud Gate* sculpture in Chicago is a good example. It reflects back to us our feeling of modernity, of being clinical, and of having conquered grime, and the dirt and messiness of life. Of being indomitable ourselves.

By solving the problem of making stainless steels hard enough for cutlery, metallurgists also unwittingly solved the problem of razors rusting, thus creating the finest cutting blade the world has ever known, and in the process altering the appearance of so many faces and bodies. Inadvertently, this domestication of shaving has also created a weapon of choice for street crime: razors that are both durable and cheap, but more than that, ultra-sharp—able to slice through several layers of leather, wool, cotton, and skin, as I knew all too well.

I weighed all this up as Brian and I talked about his new process for sharpening stainless steel razor blades. As stainless steel, a hard, tough, sharp steel impervious to water and air, has been created mostly through trial and error over the last few thousand years, it didn't seem utterly impossible that someone, even without scientific training, might yet stumble across a process for resharpening a razor blade. The microscopic world of materials is so complex and huge that only a fraction of it has been explored.

At the end of the evening, as we both left the pub, Brian shook my hand and told me he would be in touch. As he staggered off down the Dublin road bathed in the yellow light of the sodium street lamps, he turned and shouted drunkenly: "Hail to the god of steel!" I assumed he meant Hephaistos, the Greek god of metals, fire, and volcanoes, whose classical image is that of a smith at a forge. Physically handicapped, he is misshapen, suffering probably from arsenicosis, an infliction common to smiths of the time, who were exposed to high levels of arsenic poisoning during the smelting of bronze, which resulted in lameness and skin cancers. I looked back at Brian as he staggered down the road—with his walking stick and his red face—and, not for the first time that evening, wondered who he really was.

2 TRUSTED

PAPER IS SO MUCH part of our everyday lives that we may easily forget that for much of history it was rare and expensive. We wake up in the morning with paper decorating our walls, either in the form of posters and prints or as wallpaper itself. We head to the bathroom for our morning ablution and there make use of toilet paper, an item which, if absent, quickly foments a personal crisis. We head to the kitchen, where paper in the form of colored cardboard provides not just the containers for our breakfast cereals but the soundboard for them too, as they rattle their happy morning song. Our fruit juice, similarly, is contained by waxed cardboard, as is the milk. Tea leaves are encapsulated in a paper tea bag, so they can be immersed in and easily retrieved from hot water, and coffee is filtered through paper. After breakfast we may head off to face the world, but we rarely do so without taking paper with us in the form of money, notes, books, and magazines. Even if we don't leave the house with paper we quickly accrue it: we are issued paper in the form of transport tickets, we pick up a newspaper, or we buy a snack and are handed a paper receipt as a record of the purchase. Most people's work involves plenty of paperwork: despite talk of a paperless office, this has never transpired, nor does it look likely to, such is our trust in this material as a store of information.

Lunch involves paper napkins, without which personal standards of hygiene would slip dramatically. Shops are full of paper labels, without which we wouldn't know what we were buying or how much it cost. Our purchases are often contained within paper bags for their journey home. Once home we occasionally cover them in some wrapping paper as a birthday present accompanied by a paper birthday card contained in a paper envelope. Taking photos of the party, we may even print them out on photographic paper and in doing so create our material memories. Before bed we read books, blow our noses, and take one last trip to the bathroom, to convene intimately with the toilet paper again before we surrender to dreams (or perhaps nightmares of a world without paper). So what is this stuff to which we are now so accustomed?

NOTE PAPER

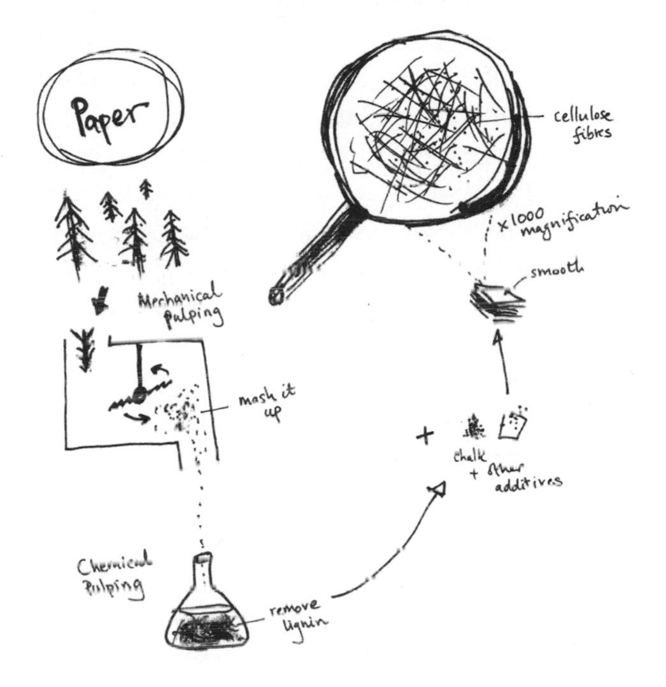

The basics of paper manufacturing, sketched out in my notebook.

Although note paper seems to be flat, smooth, continuous stuff, this is a deception: it is a mound of tiny thin fibers that resembles a bale of hay. We cannot feel its complex structure because it has been engineered at a microscopic scale that is beyond our sense of touch. We see it as smooth for the same reasons of scale that make the Earth seem perfectly round from space, while up close it has a bountiful supply of hills, valleys, and mountains.

Most paper starts out life as a tree. A tree's core strength derives from a microscopically small fiber called cellulose, which is bound together by an organic glue called lignin. This is an extremely hard and resilient composite structure that can last hundreds of years. Extracting the fibers of cellulose from the lignin is not easy. It is like trying to remove chewing gum from hair. Delignification of wood, as the process is called, involves crunching up the wood into tiny pieces and boiling them at high temperatures and pressures with a chemical cocktail that breaks down the bonds within the lignin and frees up the cellulose fibers. Once achieved, what is left is a tangle of fibers called wood pulp: in effect, liquid wood—at a microscopic scale it resembles spaghetti in a rather watery sauce. Laying this on to a flat surface and allowing it to dry yields paper.

This basic type of paper is raw and brown. Making it white, sleek, and shiny requires a chemical bleach and the addition of a fine white powder such as calcium carbonate in the form of chalk dust. Other coatings are then added to stop any ink that is laid on top of the paper from being sucked too far into the cellulose mesh, which is what causes ink to bleed. Ideally the ink should penetrate a small amount into the surface of the note paper and then dry almost instantly, depositing its cargo of colored molecules, which sit there embedded in the cellulose mesh, creating a permanent mark on the paper.

It is easy to underestimate the importance of note paper: it is a two-thousand-year-old technology, the sophistication of which is necessarily hidden from us so that, rather than being intimidated by its microscopic genius, we see only a blank page, allowing us to record on its surface whatever we choose.

PAPER RECORDS

J.[illegible]. MIODOWNIK

11,Bonython Road,

Newquay (Cornwall).

11th November,1939.

T.L.Horabin Esq.,M.P.
18,Lawrence Road,
South Norwood,
London,S.E.25.

Dear Mr.Horabin,

I have the honour to hand you copies of my application dated 22nd August,1939,and of my letter dated 8th November,1939,giving particulars referring the case I submitted to the Home Office,and I should be very obliged if you would let me know whether you could urge the Department's decision. I would not trouble you but the situation in Belgium seemed to become more precarious every day,and my wife and I are very anxious to be joined with our son who,aged nine years,could be brought over from Brussels to this country by a friend just travelling the same route.

Therefore we should be extremely obliged if you could make it possible to help us in reaching a decision at an early date.

Thanking you in advance for your trouble and kindness and anxiously awaiting your esteemed reply,

Yours sincerely,

Ismar Miodownik

A copy of a letter sent by my grandfather, Ismar Miodownik, to the British Home Office after the outbreak of the Second World War.

My grandfather's tales from when he lived in Germany at the outbreak of the Second World War absorbed me as a child, but now that he is gone, the documents he left behind must tell the stories for themselves. There is nothing quite like holding a real piece of history in your hand, such as this letter that he wrote to the British Home Office in an attempt to extract my father from Belgium, fearing a German invasion.

Paper yellows with age for two reasons. If it is made from cheap, low-grade mechanical pulp, it will still contain some lignin. Lignin reacts with oxygen in the presence of light to create chromophores (meaning, literally, "color-carriers"), which turn the paper yellow as they increase in concentration. This type of paper is used for cheap and disposable paper products, and is why newspapers yellow quickly in light.

It used to be fairly common to increase the textural quality of paper by coating it with aluminum sulfate, a chemical compound that is used primarily to purify water, but what wasn't appreciated at the time was that this treatment creates acidic conditions. This causes the cellulose fibers to react with hydrogen ions, which results in another form of yellowing. It also decreases the strength of the paper. Large numbers of books from the nineteenth and twentieth centuries were printed on this so-called acid paper and can now be easily identified in book shops and libraries by their bright yellow appearance. Even non-acid paper is susceptible to this aging, just at a slower rate.

The aging process also results in the formation of a wide range of volatile (meaning that they evaporate easily) organic molecules, which are responsible for the smell of old paper and books. Libraries are now actively researching the chemistry of book smell to see if they can use it to help them monitor and preserve large collections of books. Although it is a smell of decay, to many it is nevertheless perceived to be a pleasant one.

The yellowing and disintegration of paper are disturbing, and yet, like all antiques, paper gains an authenticity and power from its patina of age. The sensual impressions of old paper allow you to enter the past much more readily, providing a portal to that world.

PHOTOGRAPHIC PAPER

My grandfather's attempt to petition the British Home Office on behalf of his son was a success. And this was the result: my father's German identification card, which was stamped by the immigration office on his way out of Brussels on December 4, 1939. My father was nine years old at the time, and in the photo he appears oblivious to the danger of his situation. The Germans invaded in May 1940.

It is hard to overestimate the effect of photographic paper on human culture. It provided a way for identification to be standardized and verified and, in this sense, has been accepted as the final arbiter of what we look like and, by extension, who we actually are. The almost fascistic authority of the photograph originates from its (apparently) unbiased nature, which is the result of the way the image is captured. And that lies with the paper itself: since the chemicals in it record the dark and light patches of your face automatically, by simply reacting with the light reflected from it, the image itself is seen to be impartial.

The black-and-white photo of my dad started out as a white piece of paper coated in a fine gel containing silver bromide and silver chloride molecules. In 1939, when the light bouncing off my dad entered the camera lens and fell onto the photographic paper, it transformed the silver bromide and chloride molecules into little crystals of silver metal, which appear as specks of gray on the paper. If the paper had been removed from the camera at this point, the image of my dad would have been lost. This is because all the white areas where there was no image would be flooded with light, causing them instantly to react as well and creating a completely black photo. To prevent this, the photo was "fixed" in a darkroom with a chemical that washed away the unreacted silver halides from the paper. This left only the silver crystals embedded in the layer of gel on the surface of the paper. Once dried and processed it was the image of my dad that enabled him, rather than some other boy, to escape the concentration camps.

My dad is still here to tell the tale, but one day there will only be the photograph to remind us of this moment in time—a material fact of history that contributes to our collective memory. Of course, photographs are not really unbiased, but then neither are memories.

BOOKS

A photo of my bookshelf at home: books en masse are more than a library, they are a statement of identity.

The transition from an oral culture, in which knowledge was handed down through stories, songs, and apprenticeships, to a literate one, based on the written word, was held back for centuries by the lack of suitable writing material. Stone and clay tablets were used, but they were prone to fracture and were bulky and heavy to transport. Wood suffers from splitting and is susceptible to decay. Wall paintings are static and space is limited. The invention of paper, said to be one of the four great inventions of the Chinese, solved these problems, but it wasn't until the Romans replaced the scroll with the codex—or, as we call it now, the book—that the material reached its full potential. That was two thousand years ago, and it is still a dominant form of the written word.

That paper, a much softer material than either stone or wood, won out as the guardian of the written word is a remarkable materials story. It is the thinness of paper that proves to be one of its great advantages, allowing it the flexibility to survive continuous handling, but when stacked together in book form becoming stiff and strong—essentially a re-formed block of wood. With the use of hard covers to hold it all together, the book is a fortress for words for thousands of years.

The genius of this so-called codex format—a stack of papers bound to a single spine and sandwiched between covers—and the reason why it usurped the scroll, is that it allows for text on both sides of the paper and yet still provides a continuous reading experience. Other cultures achieved something similar with the concertina format—forming a stack by repeatedly folding one continuous sheet of paper in on itself—but the advantage of the codex, with its individual pages, is that many scribes could work on the same book at the same time, and after the invention of the printing press many copies of the same book could be created at the same time. As biology had already discovered, the speedy copying of information is the most effective way of preserving it.

The Bible is said to be one of the first books created in this new format, one that suited preachers of Christianity because it allowed them to locate the text relevant to their purpose using page numbers instead of laboriously searching through a whole scroll. This form of "random access memory" prefigured the digital age and may yet outlast it.

WRAPPING PAPER

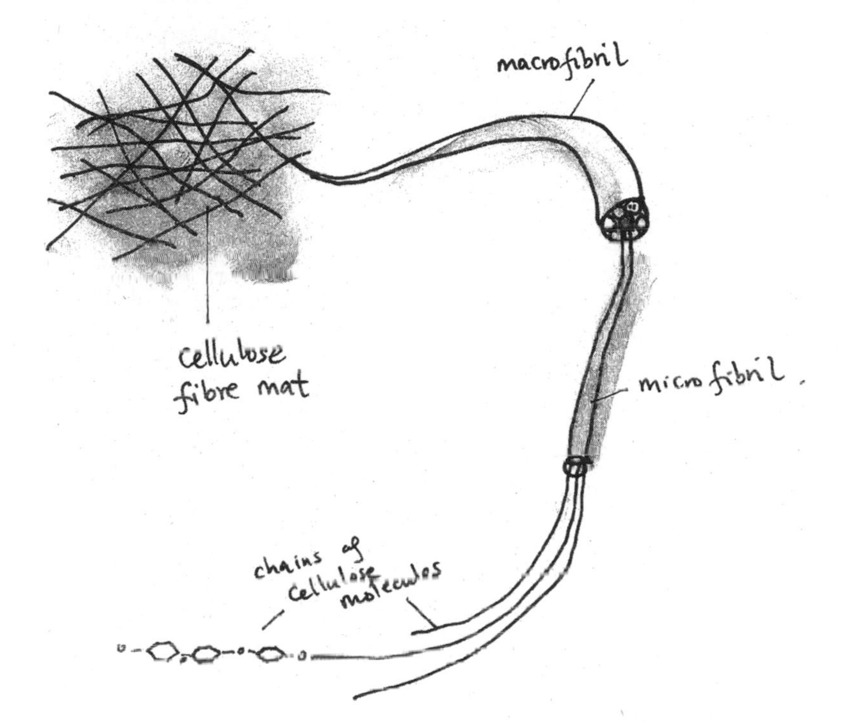

Simple paper is basically a mat of cellulose fibers.

Paper is not only useful for preserving information. In its role as a wrapping material, paper also does a good job of obscuring it. What would birthdays be like without this material, which performs the role of building excitement and anticipation better than all others? I have received presents wrapped in cloth, or hidden in a cupboard, but nothing has the magic of wrapping paper. A present really isn't a present unless it is wrapped in paper. It is the paper that, by concealing and revealing an object, ritualizes the act of giving and receiving, turning that object into a gift. This is not just a cultural association. The material has fundamental properties that make it ideal for this role.

Paper's mechanical properties lend themselves to folding and bending. The cellulose fibers of which it is made can be partially snapped in the area of maximum bend, allowing a permanent crease to form, while sufficient fibers remain intact for the material not to crack and fall apart. Indeed, in this state it pretty much maintains its ability to resist being pulled apart, but it can also be torn easily and accurately along the crease if a point of weakness—a small, initial tear—is opened up. This winning combination of mechanical properties allows it to assume the shape of any object through creasing and folding—hence the art of origami. There are very few materials as good: metal foils can hold a crease, but control of the crease is somewhat more difficult. Plastic sheeting doesn't tend to hold a crease at all, unless it is very soft, in which case it lacks the rigidity (and formality) required of a good wrapping material. So it is its ability to hold a crease while remaining stiff that makes paper uniquely suited to this purpose.

Wrapping a present with paper gives it a crispness and pristineness that emphasize the newness and value of the present inside. Paper is strong enough to protect the present when it is sent through the mail, but so weak that even a baby can rip it open. That moment of opening transports the object inside from obscurity to celebrity in a few seconds. The unwrapping of a present is akin to the act of birth; a new life for the object begins.

RECEIPTS

MARKS & SPENCER

Marks & Spencer plc
Unit 21, Building 3,
90, Southwark Street,
London SE1 0HX
Tel: 020 7633 0068
VAT No: 232 1288 92

£

00740173	S/WOLD DARK ALE	2.39DISC
[illegible]	[illegible]	[illegible]
00501105	NORFOLK BITTER	2.39DISC
00788179	BUB SWISS MILK	1.29
00047450	VEGETABLE CURRY	2.59
00100515	LAMB ROGAN JOSH	4.19
00578806	BRAEBURN BAG	2.39
00042500	*ENGLISH BUTTER	1.15
00626460	*SCOT SPRK 1.5L	0.43

Total before saving 19.21

Wines buy 3 save 10%
10% Discount 0.72-

Balance to pay 9 items 18.49

Card tendered 18.49

MASTERCARD
CARD: ************3499 EXPIRY: 04/13
REF: 11-8420/17058209/0413/1 AUTH: 228227
AID: A0000000041010 READ: ICC
Cardholder PIN Verified

06/05/11 20:00 17058209 8420 011 0134

Please retain for your records

614 095 0330 7665 5314

This is a receipt from a trip to the shop Marks & Spencer three days before my son Lazlo was born in 2011. Lazlo's mum, Ruby, had a difficult pregnancy. This was partly because she developed a craving for beer, which she would not allow herself to drink, but insisted instead that I do it for her. Some days the cravings would get so bad that, as the M&S receipt shows, I would have to drink three bottles of beer per night, with Ruby following every sip with a sometimes longing but mostly accusing gaze.

Lazlo had almost been born two weeks earlier, but in a turn of events that none of us can satisfactorily explain, he refused to emerge. After twenty-four hours we were sent home from hospital and advised that Ruby's eating hot curries might encourage Lazlo to leave the womb. Two weeks later we had got a little tired of the nightly curries that I was sent out to retrieve. I do remember liking best the lamb rogan josh, and I see we got it again that night. The logic was that the spicy diet would make life uncomfortable for Lazlo, but in truth I feel that it was we who suffered more from the digestive challenge of our extreme diet. Lazlo, by the way, is now aged two and loves spicy food.

Despite the memories of uncomfortable times the receipt evokes, I am glad I still have it. It encapsulates a different kind of intimacy from that of a photo or even a diary, where these seemingly mundane details of our lives can get lost. Sadly the receipt will not survive long enough for Lazlo to read it. It is already fading, as the thermal paper on which it is printed degrades over time. The reason for this is that printing on thermal paper does not mean adding ink to it. Rather, the ink is already encapsulated within the paper, in the form of a so-called leuco dye and an acid. The act of printing requires only a spark to heat up the paper so that the acid and dye react with each other, converting the dye from a transparent state into a dark pigment. It is this cunning paper technology that ensures that cash registers never run out of ink. But over time the pigment reverts to its transparent state and so the ink fades, taking with it the evidence of curry and beer dinners. Nevertheless, M&S enthusiastically encourages us to "Please retain for your records," something that I have dutifully done.

ENVELOPES

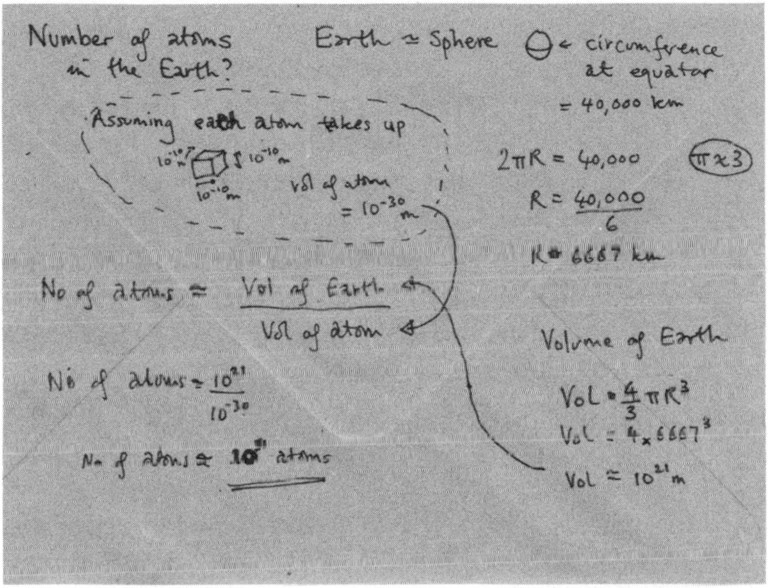

My back-of-envelope calculation of the number of atoms in the Earth, which I calculate to be 10,0000000000,0000000000,0000000000,0000000000, 0000000000 atoms.

That flash of inspiration that hits when you are on the bus or in a café requires an immediate physical expression. It needs to be noted down urgently before it is forgotten. But where? You are away from your desk and your notebooks. You search your pockets for some scrap of paper and find a letter, an electricity bill perhaps, but it will do; there is enough room on the back of the envelope for you to outline your idea. So you do, following a long line of famous scientists and engineers who have throughout history claimed the back of an envelope as an important theater of ideas.

The physicist Enrico Fermi is famous not only for solving fundamental questions of science in the restricted space on the back of an envelope but for formalizing that process. This new form of calculation—the scientific equivalent of a haiku—is called an order of magnitude calculation. This way of looking at the world prizes above all not exact answers but answers that are easily understandable and say something fundamental about the world using only the information available on a bus. They must be accurate by "an order of magnitude," which is to say that they should be correct within a factor of two or three (i.e., the true value could be as little as a third or as much as three times the result, but no more or less). Such calculations are pretty approximate but they were used by Fermi and others to demonstrate a paradox: the vast number of stars and planets in the universe should provide abundant opportunities for other intelligent life to form and therefore a great likelihood of our having encountered it, and yet, given that we have not encountered it, that same vast number is precisely what shows how rare intelligent life is.

As a kid I was so obsessed with stories of famous scientists who solved fundamental problems on the back of envelopes that I used to bring old envelopes with me to school and practice solving problems on the back of them. It was a kind of martial art for the mind, requiring only a pen and envelope. It helped me not just to clarify my thoughts but also to pass exams. The first question on my Oxford University entrance physics exam was: "Estimate the number of atoms on the Earth." I smiled when I saw this. It was classic back-of-the-envelope territory. I can't remember how I solved it in the exam, but on the previous page is my current version of this calculation.

It is the sheer ubiquity of the paper envelope that makes it so useful as a theater of ideas: even if you haven't got a notebook to sketch out your ideas, or any money to buy one, a free envelope is sure to drop through your letterbox sooner rather than later.

PAPER BAGS

There is a particular nervousness I feel when buying expensive clothes. They look and feel alien when I try them on in the shop, and no matter how many smiles and nods of approval the shop assistants give me I am never certain whether I should spend the money. When I say yes, though, I am rewarded by something of which I never tire.

It comes out first in its flat-pack condition, but then its bottom is pushed out and it makes that glorious sound of thunder as the concertinaed paper sides are deployed into their upright positions. There it sits on the shop counter, like a butterfly recently emerged from its chrysalis: perfect, elegant and poised. Suddenly my purchase seems right, now that the clothes have been allocated this special receptacle to chaperone them back home.

In stark contrast to toilet paper, in this guise paper is a refined and stylish material: light, stiff, and strong. But the strength is an illusion. The cellulose fibers that make up the paper bag are no longer accompanied by the lignin that glued them together when they were part of a tree. Although hydrogen bonds form between the fibers during the drying stage, which give the paper some strength, they have to be reinforced with synthetic adhesives. Even then it is a weak material with little water resistance: once wet, the fibers lose their hydrogen bonds, and the paper bag disintegrates fast.

It is perhaps the very fragility of paper bags that makes them appealing for their task. Expensive clothes tend to be light and fragile, and maybe the fact that paper is all that is needed for their journey home reinforces this. Paper also has high cultural status: it speaks of the craft of making, of the handmade object, associations that fit with tailor-made clothes. Again, this is an illusion in the case of paper: it is itself a fully fledged industrial product, and quite an environmentally costly one too. The impact in terms of energy usage of a single-use paper bag has been found to be greater than that of a plastic bag. So they are an indulgence, one designed to celebrate your triumphal purchases; they mark the moment of arriving home by banging inevitably into the door frames as you struggle down corridors, their soundtrack of soft rumbling thunder filling you with excitement and pride.

GLOSSY PAPER

The look and feel of paper turn out to be of the utmost importance and the secret to why it is so useful as a material. It can be transformed from rustic to official, from retro to glamorous, simply by changing the surface layer. Controlling these aesthetic considerations is vital to the economic fortunes of commercial publications.

The science of this transformation is a highly advanced topic of research. The shininess, smoothness, and weight of the paper have all been shown to be crucial to the success of certain magazines, but less appreciated is the importance of stiffness—or rather, the ease with which the paper will bend: too bendy and the paper gives an impression of cheapness; too stiff and it seems self-important. This stiffness is controlled by the addition of "sizings," — fine powder additives, such as kaolin and calcium carbonate, that among other things reduce the paper's ability to absorb moisture, causing inks to dry on its surface rather than infuse its fibers, while also allowing the whiteness of the paper to be controlled. These powders, and the binders that bond them to the cellulose fibers of the paper, create what's known as a "composite matrix." (A familiar example of a composite material is concrete, which is similarly composed of two distinct materials: cement, which is the matrix or "binder," and aggregate, which is known as a "reinforcement.") Control of this matrix determines the weight, strength, and stiffness of the paper.

It is not without problems though. It turns out that the look and feel of popular glamour magazines require a combination of stiffness and low weight, which turns the paper into a cutting instrument. The paper is so thin that its edges have the sharpness of a razor. Under most circumstances it bends rather than cuts, but if you run your fingers down one sheet at the right angle it produces a paper cut. These cuts are notoriously painful, but it is not quite clear why. It may be because they tend to occur on the fingers, which have a high density of sensory receptors, so they seem more painful than cuts elsewhere on the body. Of course, it's a price worth paying, or so the millions of people who buy glossy magazines every week seem to think.

TICKETS

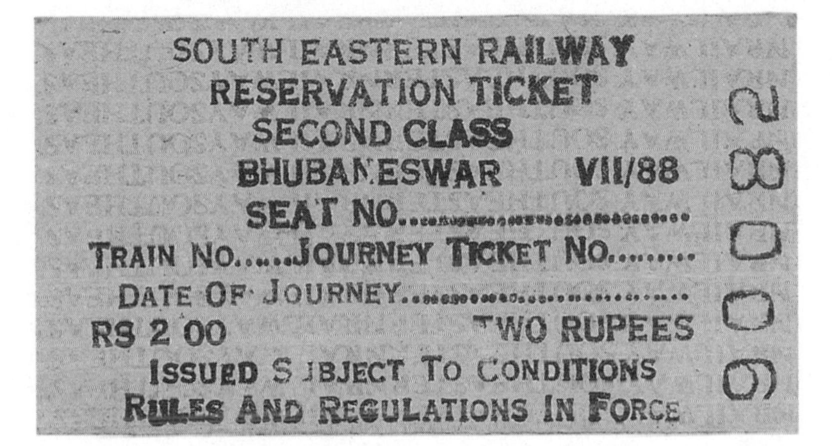

My train ticket to Bhubaneswar, a journey I took while traveling in India in 1989 with Emma Westlake and Jackie Heath.

When the thickness of paper is increased it loses its flexibility, becoming stiffer and stiffer until at some point it is stiff enough to hold itself up and not bend under its own weight. At this point it takes on new cultural roles, one of which is the permission to travel. Bus, train, and plane tickets the world over are made from thick paper called card.

All forms of human transport are engineered to be stiff and perhaps this may be the reason why the mechanical stiffness of card is so well suited to represent travel. Bendy cars are not just unusual, they are dysfunctional, because if a car chassis is not rigid enough, then the high stresses it encounters will misalign the drive mechanism. Similarly, if a train bends too much it will jump off the track, and if airplane wings sag too much under their own weight they cease to create lift. Thus the engineering of trains, planes, and automobiles requires an almost fetishistic devotion to stiffness.

Apart from stiffness, the increased solidity and strength of card imbue the ticket with a sense of authority. It is, after all, a type of temporary passport that grants a right of passage. These days a ticket must be inspected by machines as well as humans, and so it is important that the ticket is strong enough not to become bent and crumpled while being handled, stuffed in pockets, and slid in and out of wallets.

The world of travel is dominated by stiff, hard machines, and card reflects that back to us. Funnily enough, as cars and airplanes have gotten lighter and more efficient, so tickets have mirrored this, becoming thinner and thinner. Soon they will probably disappear altogether, becoming part of our digital lives.

BANKNOTES

Money is at its most seductive in its paper form. There are few more pleasing things in life than typing your PIN into a hole in the wall and being served up lovely, crisp banknotes. In sufficient quantity they are a passport to anything and anywhere in the world, and this freedom is intoxicating. They are also the most sophisticated pieces of paper that have ever been made, which they need to be because they are a literal, material manifestation of the trust we all have in the whole economic system.

To prevent forgeries the paper has a number of tricks up its sleeve. First of all, it is not made of wood cellulose, like other paper, but from cotton. This not only gives it greater strength and stops it disintegrating in the rain and washing machines, but it changes the sound of the paper: the crisp sound of paper money is one of its most notable characteristics.

It is also one of its best anti-counterfeit measures because it is hard to fake with wood-based paper. The particular texture of cotton paper is something that bank machines monitor. Humans are very sensitive to it, too. If there is any doubt about a banknote, there is an easy chemical test that can confirm whether it's cotton or not. This is done in many shops using an iodine pen. When used on cellulose-based paper, the iodine reacts with the starch in the cellulose to create a pigment and so appears black. When the same pen is used on cotton paper there is no starch for the iodine to react with, and no mark appears. This basic measure allows shops to protect themselves from counterfeits produced using color photocopiers.

But the paper has got another trick up its sleeve: watermarking. This is a pattern or picture that is embedded in the paper but can only be seen when light is transmitted through the paper—in other words, when you hold the banknote up to the light. Despite the name, this is not a water stain or an ink of any sort. It is engineered by creating small changes in the density of the cotton, so that different parts of the note look lighter and darker to produce a pattern—or, in the case of US bills greater than five dollars, the heads of presidents.

Paper money is an endangered species. Money is mostly electronic these days, and only a small percentage of transactions are carried out in cash. These are mostly low-value transactions, and electronic cash is poised to replace these too.

ELECTRONIC PAPER

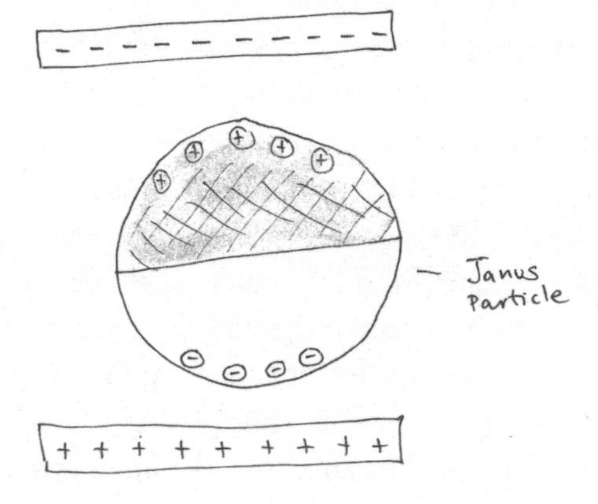

Electronic "paper" in hand-held reading devices uses electrostatic Janus particles as the electronic "ink."

Once information could be written down on paper, libraries became the most important archives of the cumulated knowledge and wisdom of a civilization. This crucial role of the library survived until very recently. Access to a great library was the key to scholarship in any university, and access to a local library was seen as a basic human right in modern societies. The digital revolution has changed the landscape considerably. Now it is possible to give everyone a complete set of the written works of the human race via a computer. But there has been a lot of resistance to the move from the physical to the digital book. Most of this is not about access but about sensual delight in the act of reading.

Suddenly, as happens so often in the history of engineering, a technology that had been around for a while but had little or no mass application came into its own. Electronic paper is a type of flat screen that displays text using real ink and is designed to be read with reflective light bouncing off it in the same way as a physical book. The difference is that electronic paper can be controlled digitally to display any text required almost instantly. When integrated with a computer chip it can store and display millions of books.

The technology relies on the ink being made into a form of the so-called Janus particle. Each particle of ink is dyed so that it is dark on one side and white on the other. The two sides are given opposite electric charges, and so each pixel on the electronic paper can be made dark or white by applying the appropriate electric charge. They are named Janus particles after the Roman god of transitions, who is depicted as having two faces and is often associated with doors and gates. Because the Janus particles are physical ink and need to physically rotate when the text is changed, they cannot be switched as rapidly as the liquid crystal display of an iPad or smartphone, and so at the moment they are unable to show movies and other snazzy stuff. They have a pleasing retro quality, which perhaps suits the written word.

The Janus particle has made reading e-books much more like the experience of reading a physical book, at least in terms of the appearance of the words on the page. It could yet be the future of the written word. However, it is unlikely that electronic paper will completely supplant books while it lacks paper's distinctive smell, feel, and sound, since it is this multisensual physicality of reading that is one of its great attractions. People love books, more perhaps than they love the written word. They use them as a way to define who they are and to provide physical evidence of their values. Books on shelves and on tables are a kind of internal marketing exercise, reminding us who we are and who we want to be. We are physical beings so it perhaps makes sense for us to identify and express our values using physical objects, which we like to touch and smell as well as read.

NEWSPAPERS

YORKSHIRE

Telegraph & Star

WAR SPECIAL

No 10,018. SHEFFIELD, MONDAY, NOVEMBER 11, 1918. ONE PENNY

THE WORLD WAR AT AN END.

STATEMENT by the PREMIER

Historic Document Signed at 5 o'clock This Morning.

HOSTILITIES ENDED AT 11 a.m. TO-DAY

OVERTHROW.

German Republics Proclaimed.

PEOPLE CONTROL.

Troops Go Over to the New Authority.

FIGHTING IN BERLIN.

Soviets Formed Everywhere.

Fellow Citizens,—The former Chancellor, Prince Max of Baden, has transferred to me the execution of the Chancellor's affairs, with the consent of all the Secretaries of State. I am planning to form a new Government in agreement with the Parties, and I shall shortly make a report to the public concerning the result. The new Government will be a People's Government. Its endeavour will necessarily be to bring peace as speedily as possible to the German people, and to fortify the freedom which they have won.

I appeal to you all for your support in the hard work that awaits us.

There is something about a printed photograph or newspaper headline that makes the event it describes more real than in any other form of news reporting. Perhaps this is because there is an undeniable reality to the newspaper itself: it is a real material object. That authenticity rubs off on the news. It can be pointed to, underlined, cut out, pinned on notice boards, stuck in a scrapbook, or archived in libraries. The news becomes an artifact, frozen in time; the event may be long gone, but it lives on as an indisputable fact because of its material presence—even if it is untrue.

In contrast, news websites seem ephemeral. Although they too are archived, there is no unique physical component to point to as evidence of the information they convey. For this reason, there is a sense in which they can be more easily manipulated, and that history itself could be altered. At the same time, it is precisely this immediacy and fluidity of content that makes the digital media so exciting. The news website is in tune with an age that sees history as much less monolithic than previous eras once did. Digital news websites are potentially much more democratic, too, for while a physical newspaper requires huge printing presses and a distribution network linking trains, planes, trucks, shops, and ultimately newspaper sellers, in the digital world a single person can communicate with the whole world with the aid of a single computer and without requiring a single tree to be cut down.

The move away from printed newspapers will change not just the internal dialogue of countries and cities but social habits too. The rustle of the paper will no longer be part of the ritual of Sunday afternoons; newspaper will no longer sit underneath muddy boots, or lounge folded up on train station benches; it will no longer protect floors from paint drips, or be wrapped around precious objects to protect them; it will no longer be crumpled into a ball, to light a fire, or be thrown cheekily at an unsuspecting sibling. None of these uses of newspaper are essential in themselves, but taken as a whole they paint a picture of a very domestic, useful, and much loved material. A material that will be missed.

LOVE LETTERS

Do you remember
the first cold night we met
when you were wearing a beard
and that lumpy brown cardigan
and I was in my fake leopardskin coat
and I asked you too many questions
and I wanted to impress you
because you felt so right
and you and the wine made me bold
and I said we should see each other again
I'd rehearsed it in my head
as we sat talking
and you said yes
and I walked away glowing
and grinning
and the next time I saw you
and we were at that strange party
where you talked to a man
in a bow tie
and I was coming down with flu
and we left in the freezing fog
and that Russian bar was closed
and we got the night bus
or was it a taxi
to your flat
where earlier we'd had a cocktail
and you lit a fire and made
hot toddies.
and we sat on the floor and kissed
and I stayed the night
and you lent me your Kurasawa t shirt
and I kept my leggings on
and in the morning we met Buzz
and had coffee together
and that was the beginning
of this most precious part of my life
and every day I think to myself
how incredibly lucky I am to have met you
and how exciting our future seems
and how full of love
and possibility.

I miss you, and it's
cold, and I'm wearing
your brown cardigan.
XXXR

Letter from my love.

Despite the march of digital technologies, it is hard to believe that paper will completely disappear as a means of communication. For some messages we trust it above all other media. There is nothing that quite grips the stomach while simultaneously making your heart skip than a letter from your beloved arriving by post. Phone calls are fine and intimate, text messages and e-mails are instantaneous and gratifying, but to hold in your hands the very material that your beloved touched and to breathe in their sweetness from the paper is truly the stuff of love.

It is a communication of more than words. There is a permanence, a physical solidity to soothe those of an insecure nature. It can be read and reread over and over again. It physically takes up space in your life. The paper itself becomes a simulacrum of the loved one's skin, it smells of their scent, and their writing is as much an expression of their unique nature as a fingerprint. A love letter is not faked, and is not cut and pasted.

What is it about paper that allows words to be expressed that might otherwise be kept secret? They are written in a private moment, and as such, paper lends itself to sensual love—the act of writing being one fundamentally of touch, of flow, of flourish, of sweet asides and little sketches, an individuality that is free from the mechanics of a keyboard. The ink becomes a kind of blood that demands honesty and expression, it pours on to the page, allowing thoughts to flow.

Letters make splitting up harder too, since like photographs they echo forever on the page. For one whose heart is broken this is a cruelty, and for those who have moved on it is a stinging rebuke of infidelity or, at the very least, a thorn of inconstancy in the side of their constructed personality. Paper, though, as a carbon-based material, has a bright solution for those wanting to be released from such a torture: a match.

3 FUNDAMENTAL

ONE SPRING DAY IN 2009, I was on my way to pick up a loaf of bread from the market when I turned the corner to find that the Southwark Towers was gone. All twenty-five floors of a classic 1970s office block had been demolished. I racked my brain to remember when I had last seen it. Surely it was last week when I had nipped out to make this exact same journey? I came over a little queasy: was I losing my grip, or had building demolition got a lot more efficient? Either way I felt less sure of myself, a little less significant. I had liked the Southwark Towers, which had automatic doors that were made when such things were groovy. Now the high-rise was gone, leaving a bigger hole in the street and my life than I was expecting; nothing looked quite the same. I made my way to the brightly colored hoardings that now surrounded the empty space it had left behind.

On the side of the hoardings, a notice announced that Europe's tallest building, the Shard, was to be built here. It had a picture of a giant pointy glass skyscraper that would rise from the ashes of the Southwark Towers over London Bridge station: the text extolled a vision of this new construction dominating the London skyline for decades to come.

I felt annoyed and worried. What if this giant glass phallus became a target for terrorists? What if it was attacked like the Twin Towers and collapsed, killing me and my family? I consulted Google Maps and reassured myself that even if this 330-meter building were to tip over sideways it would not reach my flat. At a stretch it might reach the Shakespeare Tavern, a pub nearby, but one I didn't frequent. Nevertheless, there was the suffocating dust cloud it would create, I muttered as I walked in an apocalyptic mood to get bread.

Over the next few years, I would watch this huge skyscraper being constructed on my doorstep. I would witness extraordinary sights and amazing feats of engineering, but mostly I would get to know concrete very, very well.

They started by digging a huge hole. And when I say huge, I mean *enormous.* Week after week on my bread-collecting mission, I would peer through the viewing window in the side of the hoardings, to see the progress of gigantic machines that were scooping out the dirt, digging ever deeper, as if they were mining for something. But what they were digging out was clay—clay that had been deposited there for hundreds of thousands of years by the Thames River. It was the same thick clay that has always been used to fire the bricks to make the houses and warehouses from which the city of London is built. But this clay was not going to be used to build the Shard.

One day, once all of this clay had been removed, they poured seven hundred truckloads of concrete into the hole. This would make the foundations that would hold up the enormous skyscraper and prevent the seventy-two floors above, and the twenty thousand people who would inhabit them, from sinking into the clay. They filled the enormous hole with concrete, layer after layer, building up floor after subterranean floor, until there was no gigantic hole anymore, just an underground cathedral of poured concrete, which was now slowly becoming fully solid. It was nicely done and impressively speedy, which was important because for cost reasons they had already started to build the tower before they finished its foundations.

"How long will the concrete take to dry, do you think?" a man with a dog asked me, while we were both peering through the viewing window of the hoardings. "I dunno," I lied.

My lie was intended to cut the conversation short, which it did. It was a habitual lie, born of living in London and finding ways of politely avoiding talking to strangers. Especially as I didn't know how he, or his dog, would take to my beginning our acquaintance by correcting him: concrete doesn't dry out. Quite the opposite, water is an ingredient of concrete. When concrete sets, it is reacting with the water, initiating a chain of chemical reactions to form a complex microstructure deep within the material, so that this material, despite having a lot of water locked up inside it, is not just dry but waterproof.

The setting of concrete is, at its heart, an ingenious piece of chemistry, which has powdered rock as its active ingredient. Not every type of rock will work. If you want to make your own concrete you need some calcium carbonate, which is the main constituent of limestone, a rock formed from the compressed layers of living organisms over millions of years and then fused together by the heat and pressure of the movement of the Earth's crust. You also need some rock containing silicate—silicate being a compound containing silicon and oxygen, and constituting roughly 90 percent of the Earth's crust—for which some form of clay will do. Grinding these ingredients up and mixing them together with water won't get you anywhere, unless you want to create a sludgy mud. In order to create within them the essential ingredient that will react with the water, you need to free them from their current chemical bonds.

This is not easy. These bonds are extremely stable, which is why rocks do not easily dissolve or react with many things—on the contrary, they last for millions of years. The trick is to heat them to a temperature of about 1450°C. This is a temperature far exceeding that of an average wood or coal fire, which is between 600 and 800°C if it is glowing red or yellow hot. At 1450°C a fire will glow white hot, with no tinge of red or even yellow in the flames but instead a hint of blue. It is so bright it is unnerving and almost painful to look at.

At these temperatures, rock starts to fall apart and re-form to create a family of compounds called calcium silicates. It's a family because there are lots of minor impurities that can change the outcome of what you get. To make concrete, aluminum- and iron-rich rocks are the magic ingredients, but only in the correct proportions. Once it has all cooled down, the result is a powder the gray-white color of the moon. If you put your hands through it you find that it has the silky texture of ash—there is something atavistic about it—but your hands soon feel dried out as if under a subtle type of attack. This is a very special material with a very dull name: cement.

If you now add water to this powder it sucks it up with ease and darkens. But instead of forming a slushy mud, which is what happens if you add water to most powdered rock, a series of chemical reactions takes place to form a gel. Gels are semisolid and wobbly types of matter—the jelly served at children's parties is a gel, and so too is a lot of toothpaste. It doesn't slosh around like a liquid because it has an internal skeleton that prevents the liquid moving. In the case of jelly this is created by the gelatin. In the case of cement, the skeleton is made up of calcium silicate hydrate fibrils, which are crystal-like entities that grow from the calcium and silicate molecules, now dissolved in the water, in a way that appears almost organic (see the picture below). So the gel that forms inside cement is constantly changing as the solid internal skeleton grows and further chemical reactions take place.

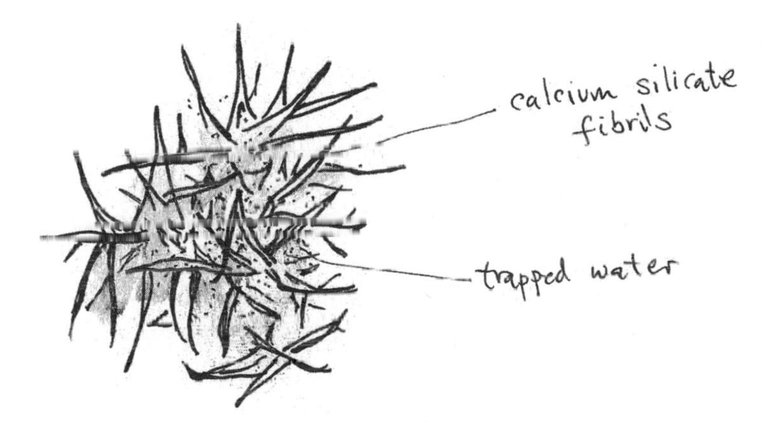

A sketch of calcium silicate fibrils growing inside the setting cement.

As the fibrils grow and meet, they mesh together, forming bonds and locking in more and more of the water, until the whole mass transforms from a gel to a solid rock. These fibrils will bond not only to each other but also to other rocks and stones, and this is how cement turns into concrete. Cement is used to bond together bricks to make houses and stones to make monuments, but in both these cases it is wedged between the cracks as the minority component, an urban glue. When it is made into concrete by mixing it with small stones, which play the role of tiny bricks, it fulfils its potential to become a structural material.

As with any chemical reaction, if you get the ratio of the ingredients wrong, then you get a mess. In the case of concrete, if you add too much water there won't be enough calcium silicate from the cement powder to react with, and so water will be left over within the structure, which makes it weak. Similarly, if you add too little water there will be unreacted cement left over, which again weakens the structure. It is usually human error of this sort that proves the undoing of concrete. Such poor concrete can go undiscovered but then lead to catastrophe many years after the builders have departed. The extent of the devastation due to the 2010 earthquake in Haiti was blamed on shoddy construction and poor-quality concrete: an estimated 250,000 buildings collapsed, killing more than 300,000 people, and making a million more homeless. What is worse is that Haiti is by no means unusual. Such concrete time bombs are scattered throughout the world.

Tracking down the origin of such human errors can be tricky since, from the exterior, the concrete looks fine. The supervising engineer of the building of JFK Airport noticed through routine tests that the concrete arriving on trucks before noon had good strength when it set, but that arriving just after noon was substantially weaker. Puzzled, he investigated all possible reasons for this but was unable to find the answer until he resorted to following the truck delivering the concrete on its journey to the airport. He found that around noon the driver was in the habit of taking a break for lunch and would hose the concrete with water before doing so in the belief that adding extra water would keep the concrete liquid for longer.

While digging the foundations for the Shard and its support-

Remains of a Roman bathhouse found by the Shard engineers.

ing structures, the engineers found evidence of a type of concrete that predates the modern stuff: Roman concrete. It was holding together the remains of a Roman bathhouse they found when they demolished my local fish and chip shop, which had resided next to the now ex–Southwark Towers. The Romans got lucky with concrete. Instead of having to experiment with heating up different combinations of ground-up rock to white-hot temperatures, they found ready-made cement in a place called Pozzuoli just outside Naples.

Pozzuoli stinks—literally. It took its name from the Latin *putere* (to stink), the smell coming from the sulfur in the volcanic sands nearby. The upside of the smell was that the region had been the recipient of lava and eruptions of ash and pumice for millions of years. This volcanic ash resulted from the super-heating of silicate rock, which was then spewed out of a volcanic vent—a process suspiciously similar to that of making modern cement. All the Romans had to do was put up with the smell and mine the rock powder that had been accumulating for millions of years. This naturally made cement is slightly different from modern ("Portland") cement and requires the addition of lime to make it set. But once they had worked this out and added stones for strength, they had in their hands for the first time in human history the fundamentally unique building material that is concrete.

The composite nature of a brick building is part of its appeal. The brick itself is a unit of construction that is designed to fit in the hand, giving the whole a human scale. Concrete is fundamentally different from this building material, because it starts off as a liquid. This means that buildings made from concrete can be poured, and what is created is a continuous structure, from the foundations to the roof, without any joins.

The mantra of a concrete engineer is: you want foundations, we will pour you foundations; you want pillars, we will pour you pillars; you want a floor, we will pour you a floor; you want it twice the size?—no problem; you want it curved?—no problem. With concrete, if you can build the mold, you can create the structure. The power of the stuff is palpable, and addictive to anyone who visits the building sites where the stuff is being made. Week after week I would peer through the viewing hole on the site of the Shard and be transfixed by what I saw. A building was growing out of the foundations; it was being poured into existence by human ants. Powdered rock and stones arrived at the site and were transformed by the simple addition of water into rock again. It is a philosophy as much as it is an engineering technique, completing a cycle that starts when the Earth's mantle creates rock and stone through mountain building, which is then mined by humans and transformed back into our own artificial mountains of rock, made to our own design, where we live and work.

The existence of concrete feeds the ambition of engineers. Once the Romans had invented it, they realized that it would allow them to build the infrastructure of their empire. It allowed them to build ports wherever they wanted, because their concrete could set under water. They could build aqueducts and bridges, too—the very infrastructure needed to transport the raw ingredients for concrete to wherever they were needed, instead of relying on local stone or clay. In this sense, concrete is ideally suited to empire building. The most impressive piece of Roman concrete engineering, however, is in its capital: the dome of the Pantheon in Rome. Still standing today, it is two thousand years old but still the largest unreinforced concrete dome in the world.

Although the Pantheon survived the fall of the Roman Empire, concrete as a material did not. There were no concrete structures for more than a thousand years after the Romans stopped making it. The reason for this loss of the materials technology remains a mystery. Perhaps the material was lost because it was industrial in nature and needed an industrial empire to support it. Perhaps it was lost because it was not associated with a particular skill or craft, such as ironmongery, stonemasonry, or carpentry, and so was not handed down as a family trade. Or perhaps it was lost because Roman concrete, good as it was, did have one crucial flaw, a flaw that the Romans knew all about but could not solve.

There are two ways of breaking a material. One is to break it plastically, which is what happens when, for example, you pull a piece of chewing gum apart: the material is able to rearrange itself and so flows and gets thinner in the middle until eventually it is separated into two pieces. This is what you need to do to break most metals, but it takes a lot of energy to get metals to flow like this (because you have to move a lot of dislocations), which is why they are such strong and tough materials. The other way of breaking a material is to create a crack through it, which is how a glass or a tea cup breaks: unable to flow in order to accommodate the stress that is pulling it apart, a single weakness in this type of material compromises the integrity of the whole, and it splits or shatters. This is how concrete breaks, which was a big headache for the Romans.

The Romans never solved this problem and so only used concrete in situations where it was being compressed rather than stretched, such as in a column, dome, or the foundations of a building, where every part of the concrete was being squeezed together by the weight of the structure. Under such compression, concrete remains strong even when cracks form. If you visit the two-thousand-year-old concrete Pantheon dome, you will see that over the years it has developed cracks, perhaps as the result of earthquakes or subsidence, but these cracks do not endanger the structure because the whole dome is under compression. When the Romans tried to make suspended floors or beams of concrete, which had to withstand a bending stress, they would have found that even the slightest crack causes the structure to collapse. When the material on either side of the crack is being pulled apart by its own weight and the weight of the building, it has no way to resist. So to use concrete to its full potential, as we do today, to build walls, floors, bridges, tunnels, and dams, this problem had to be solved. This didn't happen until the European Industrial Revolution hove into view, and even then the solution came from a very unexpected source.

A Parisian gardener, Joseph Monier, had a liking for making

his own plant pots. At the time, in 1867, these were made of fired clay, meaning they were weak, brittle, and expensive to make, especially if one wanted large pots to accommodate the craze for growing tropical plants in glasshouses. Concrete seemed to offer the solution. It could be used to make huge pots much more easily than clay because it didn't need to be fired in a furnace. It was also cheaper for the same reason. But it was still weak in tension, so in the end his concrete pots cracked just the same as terra cotta.

Joseph's solution was to embed loops of steel inside the concrete. He couldn't have known that cement bonds very well to steel. It could easily have turned out that the steel was like the oil in the vinaigrette of concrete, preferring to keep to itself. But no, the calcium silicate fibrils inside concrete stick not just to stone but also to metal.

Concrete is essentially a simulacrum of stone: it is derived from it and is similar in appearance, composition, and properties. Concrete reinforced with steel is fundamentally different: there is no naturally occurring material like it. When concrete reinforced with steel comes under bending stresses, the inner skeleton of steel soaks up the stress and protects it from the formation of large cracks. It is two materials in one, and it transforms concrete from a specialist material to the most multipurpose building material of all time.

Something else Joseph couldn't have known at the time turns out to be one of the keys to the success of his reinforced concrete. Materials are not static things: they respond to their environment, and especially to temperature. Most materials expand when they get warmer and contract when they get cooler. Our buildings, roads, and bridges expand and contract like this, observing day and night temperature cycles, as if they are breathing. It is this expansion and contraction that causes a lot of the cracks in roads and buildings, and if it is not taken into account in their design, then the stresses that build up can destroy the structure. Any engineer guessing the outcome of Joseph's experiments might have assumed that concrete and steel, being so different, would expand and contract at such different rates that they would tear each other apart; that in the heat of the summer or in the depths of winter in Joseph's garden, the steel would break out of the concrete, causing the pots to rupture. Perhaps this is why it took a gardener to try the experiment at all—it just seems so obvious that it would not work. But, as luck would have it, steel and concrete have almost identical coefficients of expansion. In other words, they expand and contract at almost the same rate. This is a minor miracle, and Joseph was not the only one to notice it. An Englishman, William Wilkinson, had also happened upon this magic combination of materials. Reinforced concrete's time had come.

Go to any of the world's many developing countries and you will find that millions of the poorest people live in shantytowns built of mud, wood, and corrugated steel roofs. These dwellings are very vulnerable to the elements. They are oppressively hot in the sun and leaky and unstable in the rain. They are regularly destroyed by storms, washed away by floods, or cleared by bulldozers in the service of the police and the powerful. To build a stable defense against the elements and those who would oppress you requires a material that is not just strong but also fire-, storm-, and waterproof, and, crucially, cheap enough for everyone in the world to afford it.

Reinforced concrete is that material. At £100 per ton, concrete is by a long way the cheapest building material in the world. But it also lends itself to mechanization of construction and so allows further cost reductions. One person and a concrete mixer can build the foundations, walls, floors, and roof of a house in a few weeks. Because all of these elements are part of the same structure it can easily last a hundred years in all weathers. The foundations protect it from water infiltration and are impervious to insect and mold attack. The walls will resist collapse and support glass windows securely. It needs very little maintenance: the tiles will not be blown off it, because it has none; the roof is an integral part of the structure, and vines, plants, and even grass can be grown on top of it for sustenance and to thermally insulate the building. (The fact that such flat-roof gardens are only possible thanks to concrete's reinforcing steel internal skeleton — only domes, like the Pantheon, would be possible otherwise — is a pleasing salute to one of its gardener inventors.)

As work continued on the Shard I found I no longer needed to visit the viewing window cut into the hoarding at the site itself. In fact this afforded me the worst view. All the action was now at the top of the growing tower. I got the best view from my roof, and it soon became a habit for me to get up in the morning and contemplate the Shard's progress with my morning coffee. I started measuring how it had grown with a chalk mark against the chimney stack. It went up, up, and up! At the peak of activity, I calculated that the engineers were adding approximately a whole floor every few days.

What made this possible was that the concrete was being continually cast. It arrived by truck at the bottom of the building and was pumped up into a mold at the top. Meanwhile, the mold, which was the size and shape of a floor of the building, was fitted with steel rods that would become the internal skeleton of the concrete tower. Once a floor had been cast, it was then used to support the mold, which was moved up a story, ready for the next floor to be cast. And so the process was repeated; this building was growing. Growing, I calculated, at a rate of three meters a day.

What was more staggering to me was that this bootstrapping process could seemingly continue for as long as you cared to move the mold up another floor and pour in more concrete. It was like a bud on a growing sapling. In reality, though, there are currently limits to the process. The engineers of the Burj Khalifa in Dubai, which is almost three times taller than the Shard, found that the capacity of the machinery to pump concrete vertically to the top of that tower proved to be a severe problem.

Nevertheless, the method is ingenious. This mechanization of the process of building is what makes concrete such a modern material. It lends itself to pouring and molding, to the rapid building of vast structures. The big structures of old, such as the stone cathedrals of Europe or the Great Wall of China, took decades to build. The central core of the Shard, one of the tallest buildings in Europe, took less than six months. The material enables you to think big, to dream. It is the material that has allowed the ambition of civil engineers to be realized. The Hoover Dam is built from reinforced concrete, as are the Millau Viaduct and Spaghetti Junction.

The Millau Viaduct, in France, one of the most beautiful bridges in the world, is made of reinforced concrete.

One day, the Shard stopped growing, and then over a matter of days the paraphernalia of the concrete mold disappeared. What was left was a concrete tower seventy-two stories high: it was gray, raw, and wrinkly like a newborn. Work began at the bottom again, while the newest concrete tower in London swayed quietly in the wind, with seemingly nothing to do but watch while human ants swarmed at its base. But it was not idle. Inside the material, the fibrils of calcium silicate hydrate were growing, meshing together and bonding with the stones and steel. The tower, in doing so, was getting stronger. Although concrete reacts with water to harden to a reasonable strength within twenty-four hours, the process by which this artificial rock develops its internal architecture and so its full strength takes years. As I write this, the concrete core of the Shard continues to harden and strengthen, albeit imperceptibly.

The Shard during construction.

Once at full strength, the concrete structure will take the weight of the twenty thousand people who will be inhabiting it by day.

It will take the weight of all their thousands of desks and chairs, all thc furniture and computers, as well as tons and tons of water. It will do this day in, day out, without visibly deforming. The floors will remain rigid and solid. And it is capable of supporting the building's occupants and protecting them from the elements without complaint for thousands of years. If the concrete is looked after, that is.

Because despite reinforced concrete's impressive credentials as a building material, it does need care. In fact its vulnerability has the same origin as its strength: its internal structure.

In ordinary circumstances, exposed to the elements, the steel that is used to reinforce concrete is prone to rusting. When that steel is encased within concrete, the alkaline conditions create a layer of iron hydroxide on top of the steel, which acts as a protective skin. But during a building's lifetime, arising from normal wear and tear and the expansion and contraction that takes place during winters and summers, small cracks will appear in the concrete. These cracks can allow water inside, water that can freeze, expanding and creating a deeper crack. This type of attrition and erosion is what all stone buildings have to put up with. It is also what mountains have to put up with, which is how they get eroded. To prevent stone or concrete structures being similarly afflicted, maintenance of their fabric needs to be carried out every fifty years or so.

But concrete can suffer from a more pernicious type of damage. This occurs when lots of water gets into concrete and starts to eat away at the steel reinforcement. The rust expands inside the structure, creating further cracking, and the whole internal steel skeleton can be compromised. It is particularly likely to happen in the presence of saltwater, which destroys the iron hydroxide protection and rusts the steel aggressively. Concrete bridges and roads in cold countries, which are regularly exposed to salt (such as is used to clear snow and ice), are vulnerable to this type of chronic deterioration. Recently London's Hammersmith Flyover was shown to be suffering from concrete decay of this kind.

Given that literally half of the world's structures are made from concrete, the upkeep of concrete structures represents a huge and growing effort. To make matters more difficult, many of these structures are in environments that we don't want to have to revisit on a regular basis, such as the Øresund Bridge connecting Sweden and Denmark, or the inner core of a nuclear power station. In these situations it would be ideal to find a way to allow concrete to look after itself, to engineer concrete to be self-healing. Such a concrete does now exist, and although it is in its infancy it has already been shown to work.

The story of these self-healing concretes started when scientists began to investigate the types of life forms that can survive extreme conditions. They found a type of bacterium that lives in the bottom of highly alkaline lakes formed by volcanic activity. These lakes have pH values of between 9 and 11, which will cause burns to human skin. Previously it had been thought, not unreasonably, that no life could exist in these sulfurous ponds. But careful study revealed life to be much more tenacious than we thought. Alkaliphilic bacteria were found to be able to survive in these conditions. And it was discovered that one particular type called *B. pasteurii* could excrete the mineral calcite, a constituent of concrete. These bacteria were also found to be extremely tough and able to survive dormant, encased in rock, for decades.

Self-healing concrete has these bacteria embedded inside it along with a form of starch, which acts as food for the bacteria. Under normal circumstances these bacteria remain dormant, encased by the calcium silicate hydrate fibrils. But if a crack forms, the bacteria are released from their bonds, and in the presence of water they wake up and start to look around for food. They find the starch that has been added to the concrete, and this allows them to grow and replicate. In the process they excrete calcite, a form of calcium carbonate. This calcite bonds to the concrete and starts to build up a mineral structure that spans the crack, stopping further growth of the crack and sealing it up.

It's the sort of idea that might sound good in theory but never work in practice. But it does work. Research now shows that cracked concrete that has been prepared in this way can recover 90 percent of its strength thanks to these bacteria. This self-healing concrete is now being developed for use in real engineering structures.

Concrete cloth.

Another type of concrete with a living component is called filtercrete. This is a concrete that has very particular porosity, such that it allows naturally occurring bacteria to colonize it. The pores in the concrete also allow water to flow through it, reducing the need for drains, while the bacteria inside the concrete purify the water by decomposing oils and other contaminants.

And there is also now a textile version of concrete called concrete cloth. This material comes in a roll and needs only water to be added for it to harden into any shape you like. Although this material has great sculptural potential, perhaps its biggest application may be in disaster zones, where tents made in situ from rolls of concrete dropped from the air can create a temporary city in a matter of days, one that will keep out the rain, wind, and sun for years while rebuilding efforts continue.

What happened next at the Shard, though, was no such celebration of concrete's potential. Instead the builders were slowly but systematically cladding the outside of the Shard with steel and glass to remove all traces of its concrete core. The implication was stark: they were ashamed of concrete. It had no place in how this building was to face the outside world or its inhabitants.

This attitude is shared by most people. Concrete is perceived as fine for the construction of a motorway bridge or for a hydroelectric dam, but it is not deemed a suitable material for building within a city. The use of concrete to express a sense of freedom and liberty, as in London's Southbank Centre in the 1960s, is now unthinkable.

The 1960s were heady days for concrete. It was used boldly to reinvent city centers, to build a modern world. But somewhere along the way this association was lost, and people decided it wasn't the material of the future after all. Perhaps too many poor-quality concrete multistory car parks were built, too many people were mugged in graffiti-covered concrete underpasses, or too many families felt dehumanized by living in a concrete high-rise estate. These days concrete is regarded as necessary, cheap, functional, gray, dreary, stained, inhuman, but most of all ugly.

But the truth is that cheap design is cheap design whatever the material. Steel can be used in good or bad design, as can wood or bricks, but it is only with concrete that the epithet of "ugly" has stuck. There is nothing intrinsically poor about the aesthetics of concrete. You only have to look at the Sydney Opera House, whose iconic shell enclosures are made of concrete, or the interiors of

London's Barbican Centre to realize that the material is capable of—and in fact makes possible—the greatest and most extraordinary architecture. This has not changed since the 1960s. It is the look of concrete that is now felt to be unacceptable, which means that concrete is now routinely hidden away from sight, providing the core and foundations but not allowed to be visible.

Many new versions of concrete have been invented to refresh its aesthetic appeal. The latest is self-cleaning concrete, which contains titanium dioxide particles. These sit on its surface but are microscopic and transparent, so it looks no different. However, when they absorb UV light from the sun, the particles create free radical ions, which break down any organic dirt that comes into contact

Dives in Misericordia Church in Rome.

with them. The remains are washed away by the rain or blown away by the wind. A church in Rome called Dives in Misericordia has been constructed with such self-cleaning concrete.

In fact, the titanium dioxide does more than clean the concrete: it can also reduce the level of nitrogen oxide in the air, produced by cars, like a catalytic converter. Several studies have shown that this works, and open up the possibility that in the future buildings and roads may not be purely passive; they may purify the air much like plants.

Now that the Shard is complete, all the concrete is hidden from sight, encased in more acceptable materials. But our ugly secret, and that of the Shard, is that concrete is literally the foundation of our whole society: it is the basis of our cities, our roads, our bridges, our power stations—it is 50 percent of everything we make. But like bone we prefer it on the inside; when it sticks out we are repulsed. This may not be a permanent situation. Maybe it is just the end of the second wave of enthusiasm for concrete. The first was started by the Romans and ended for mysterious reasons. The new concrete that is coming along is more sophisticated and may yet reverse our tastes again, igniting a third wave of enthusiasm, this time for smart concrete with embedded bacteria that allow it to create living, breathing architecture, thus changing our relationship with this most fundamental of materials.

TAKE A PIECE OF dark chocolate and pop it in your mouth. For a few moments you will feel its hard corners against your palate and tongue but taste little in the way of flavor. It is almost impossible to resist the urge to give it a good bite, but try very hard not to, so that you can experience what happens next: the lump becoming suddenly limp as it absorbs the heat from your tongue. As it becomes liquid, you will notice your tongue feels cooler, and then a combination of sweet and bitter flavors floods your mouth. These are followed by fruity and nutty sensations, and finally an earthy, muddy taste down the back of your throat. For one blissful moment you will be in thrall to the most deliciously engineered material on earth.

Chocolate is designed to transform into a liquid as soon as it hits your mouth. This trick is the culmination of hundreds of years of culinary and engineering effort, aimed initially at creating a popular drink that could hold its own against tea and coffee. That effort failed miserably until chocolate manufacturers realized that making hot chocolate in the mouth instead of in a saucepan was much more delightful, much more modern, and far more widely liked. In effect they created a solid drink, made possible by their understanding and control of crystals—specifically, cocoa butter crystals. The chocolate industry has never looked back.

Cocoa butter is one of the finest fats in the vegetable kingdom, slugging it out with dairy butter and olive oil for pole position. In its pure form it looks like fine unsalted butter and is the basis not just of chocolate but of luxury face creams and lotions. Don't let this put you off—fats have always provided humans with more than just food, in the form of candles, creams, oil lamps, polish, and soap. But cocoa butter is a special fat for many reasons. For one, it melts at body temperature, meaning that it can be stored as a solid but becomes a liquid when it comes into contact with the human body. This makes it ideal for lotions. Moreover, it contains natural antioxidants, which prevent rancidity, so it can be stored for years without going off (compare that to butter made from milk, which has a shelf life of only a few weeks). This is good news for face cream makers but also for chocolate manufacturers.

Cocoa fat has another trick up its sleeve: it forms crystals, and these are what give chocolate bars their mechanical strength. The major component of cocoa butter is a large molecule called a triglyceride, which forms crystals in many different ways, depending on how these triglycerides are stacked together. It's a bit like packing the trunk of a car: there are many ways to do it, but some take up more space than others. The more tightly packed the triglycerides, the more compact the crystals of cocoa fat. And the denser the cocoa fat, the higher its melting point and the more stable and stronger it is. These denser forms of cocoa are also the hardest to make.

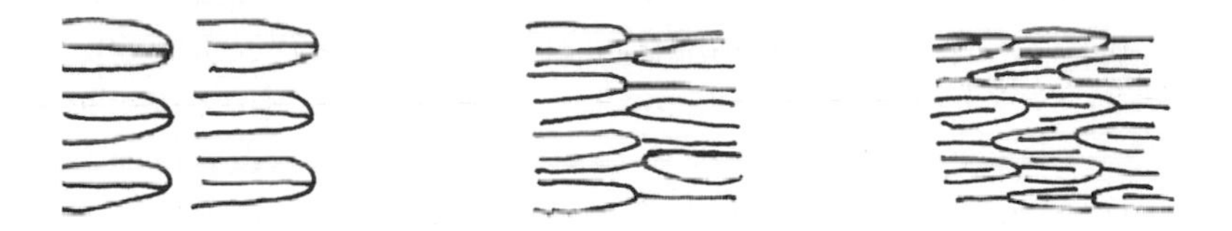

Sketch illustrating different ways to pack triglyceride molecules into a crystal form, each way having a different structure and density.

Types I and II crystals, as they are called, are mechanically soft and quite unstable. They will, if given any chance at all, transform into the denser Types III and IV. Nevertheless they are useful for making chocolate coatings on ice creams, because their low melting point of 16°C allows them to melt in the mouth even when cooled by the ice cream.

Types III and IV crystals are soft and crumbly and have no brittle "snap" when broken. The mechanical property of the snap is important to chocolatiers because it adds surprise and drama to our experience of the chocolate. For example, it allows them to create hard outer shells with which to encase soft centers, providing a textural contrast. From a psycho-physics perspective, meanwhile, brittleness and the sound associated with cracking open a chocolate are linked with freshness, which again adds to the enjoyment of eating chocolate with a "snap." Anyone who has tucked into a bar of chocolate expecting it to be hard and brittle only to find it gooey and melted knows just how disappointing losing the snap can be. (Although it is fair to say that gooey chocolate has its place as well . . .)

For all these reasons chocolate makers tend to want to avoid Types III and IV crystals, but unfortunately they are the easiest to create: if you melt some chocolate and then let it cool down, you will almost certainly form Types III and IV crystals—this chocolate feels soft to the touch, has a matte finish, and melts easily in the hand. These crystals will transform into the more stable Type V over time, but on the way they will eject some sugar and fat, which will appear as white powder on the surface of the chocolate—called bloom.

Chocolate showing fat bloom.

Type V is an extremely dense fat crystal. It gives the chocolate a hard, glossy surface with an almost mirror-like finish, and a pleasing "snap" when broken. It has a higher melting point than the other crystal types, melting at 34°C, and so only melts in your mouth. Because of these attributes, the aim of most chocolatiers is to make Type V cocoa butter crystals. This is easier said than done. They have to be created through a process called tempering, in which preformed "seed" Type V crystals are added during the final process of solidification. These give the slower-growing Type V crystals a head start over the faster-growing Type III and IV crystals, allowing the whole liquid mass to solidify into the denser form of crystal structure before the Type III and IV crystals have a chance to get going.

When you put pure dark chocolate into your mouth and sense it start to liquefy, what you are feeling are the Type V cocoa butter crystals that are holding the chocolate together starting to wobble. If they have been cared for properly, they will have spent their entire life at temperatures below 18°C. Now, in your mouth, they experience higher temperatures for the first time. This is the moment they have been created for. It is their first and last performance. As they warm up and reach the threshold of 34°C they start to melt.

This change from solid to liquid—a so-called transformation of state—requires energy to break the atomic bonds that are holding the molecules of a crystal together, thus freeing them to move around as a liquid. So as the chocolate reaches its melting point, it takes this extra energy that it needs from your body. The chocolate gets this energy in the form of latent heat, as it is called, from your tongue. You perceive it as a pleasant cooling effect, similar almost to sucking a mint. It's the same cooling effect that is produced when you sweat, but rather than a solid becoming a liquid, instead a liquid (your sweat) changes state into a gas, absorbing the latent heat required to do so from your skin. Plants use the same process to cool themselves down.

In the case of cocoa crystals, the coolness of the chocolate melting is accompanied by the sudden production of a warm thick liquid in the mouth, and it is this wild combination of impressions that is responsible for the unique feel of chocolate in the mouth—it is the beginning of the hot chocolate experience.

What happens next is that the ingredients of the chocolate, once bound together by the rigid cocoa butter matrix, are now free to flow to your taste buds. The grains of the cocoa nut, which were once encapsulated in the solid cocoa butter, are now released. Dark chocolate usually contains 50 percent cocoa fat and 20 percent cocoa nut powder (referred to as "70 percent cocoa solids" on the packaging). Almost all the rest is sugar. Thirty percent sugar is a lot. It's the equivalent of putting a spoonful of sugar in your mouth. Nevertheless dark chocolate isn't overly sweet; sometimes it's not sweet at all. This is because at the same time that the sugars are released by the melting cocoa butter, so are chemicals known as alkaloids and phenolics from the cocoa powder. These are molecules such as caffeine and theobromine, which are extremely bitter and astringent. They activate the bitter and sour taste receptors and complement the sweetness of the sugar. Balancing these basic tastes to give the chocolate a rounded flavor is the first task of the chocolatier. The addition of salt as a flavor enhancer, as well as adding another dimension to modern chocolates, has in turn led to chocolate being used as an ingredient in savory dishes: it is the basis of the Mexican dish *pollo con mole,* which is chicken cooked in dark chocolate.

However, cooked chocolate tastes different from eating chocolate for another reason. Although basic taste is generated on the tongue by the taste buds, which distinguish between bitter, sweet, salty, sour, and umami (meaty or savory), most flavor is experienced through smell. It is the smell of chocolate from within your own mouth that is responsible for its complex taste. When you cook chocolate, many of its flavor molecules evaporate or are destroyed by the cooking. This is a problem not just for hot chocolate but also for coffee and tea. It is why you need to drink those drinks within minutes of being brewed, otherwise the flavor disappears into the air. It is also why you lose much of your sense of taste when you have a cold—because the smell receptors in the nose are covered by mucus. The genius of creating hot chocolate in the mouth is that the cocoa butter encapsulates the flavor molecules until the moment you eat it, and only then does it release its cocktail of more than six hundred exotic molecules into your mouth and up your nose.

Some of the first flavors that you detect up your nose are fruity ones belonging to the ester family of molecules. These molecules are responsible for the ripe smell of beer, wine, and, more obviously, fruits. But these esters are not present in the raw cocoa bean. I know this because I have eaten a raw cocoa bean and it tastes horrible: it is fibrous, woody, bitter, and bland; there is no fruitiness, no hint of a chocolate taste, and certainly no reason to taste one again. It takes quite a bit of engineering to turn these rather exotic-looking but dull-tasting beans into chocolate. So much so, in fact, that it gets you wondering how it was ever invented at all.

A cocoa tree laden with cocoa pods.

Cocoa trees grow in tropical climates and produce fruit in the form of large fleshy cocoa pods. These look like some form of wild and leathery orange or purple melon. The pods grow directly out of the trunk of the tree, rather than from a branch – making them seem suspiciously unevolved and prehistoric. You can imagine dinosaurs trying to eat them (and spitting them out).

Inside each pod there are thirty to forty soft, white, fat almond-shaped seeds the size of small plums. On my first encounter with these cocoa nuts I popped one in my mouth and chewed it excitedly. And then I spat it out as soon as I realized what it tasted like. I wondered if it really was a cocoa nut, and was told that it was. "But it doesn't taste of chocolate," I complained, dripping with sweat. I was at the time helping to pick the cocoa pods on a Honduran cocoa plantation while simultaneously being attacked by mosquitoes. Disappointment and discomfort notwithstanding, I realized I was being petulant, and that I must have sounded like one of the gold ticket holders from Roald Dahl's book *Charlie and the Chocolate Factory.* The setting, too, was almost fictionally exotic: the small gnarled cocoa trees were growing in the shade of banana and coconut trees, stuffed with fruit, whose leaves filtered the bright sun into a thousand shades of green. What happened next was straight out of the Willy Wonka school of chocolate making too: we harvested the cocoa beans using machetes, and then deposited them in a heap on the ground, where we left them to rot.

I found out later that this is not an eccentric custom of Honduran cocoa farmers; it is how all chocolate is made. Over two weeks the heaps of beans start to decompose and ferment, and in the process they heat up. This serves the purpose of "killing" the cocoa seeds, inasmuch as it stops them from germinating into cocoa plants. But more importantly it chemically transforms the raw ingredients of the cocoa bean into the precursors of the chocolate flavors. If this step doesn't take place, whatever else you do, you won't get anything remotely like chocolate.

It is during fermentation that the fruity ester molecules are created, the result of a reaction between the alcohols and the acids that are created by enzymes acting within the cocoa beans. As with all chemical reactions there are a vast number of different variables that affect this outcome: the ratio of the ingredients, the surrounding temperature, the availability of oxygen, and many others. This means that the taste of chocolate is highly dependent not just on the ripeness and species of the cocoa bean, but also on how high the rotting piles of beans are stacked, how long they are left to rot, and generally what the weather is like.

If all this makes you wonder why chocolate makers rarely talk about these subtleties, it is because they are a secret. On the face of it cocoa seems to be like other commodities: a basic ingredient, like sugar, that is bought and sold on world markets, fueling a billion-dollar industry in edible products. But what is much less talked about is that, just like coffee and tea, different varieties of bean and different techniques of preparation create vastly different tastes. A detailed understanding of both is required to buy the right beans, and when it comes to creating the finest chocolates this knowledge is closely guarded. Controlling quality, meanwhile, also means taking into account the variability of tropical weather and the sporadic influx of disease. All in all, producing quality chocolate requires a huge amount of care and attention, which is why good dark chocolate is expensive.

What you get for your money, though, is not just the delicate fruity flavors from the fermented esters, but a set of earthy, nutty, almost meaty flavors. These are produced in the process that comes after fermentation, when the beans are dried and roasted. As with coffee making, roasting turns each bean into a mini chemical factory, in which a new set of reactions takes place. First, the carbohydrates within the bean, which are mostly sugar and starch molecules, start to fall apart because of the heat. This is essentially the same thing that happens if you heat up sugar in a pan: it caramelizes. Only in this case the caramelizing reaction takes place inside the cocoa bean, turning it from white to brown and creating a wonderful range of nutty caramel flavor molecules.

The reason why any sugar molecule—whether in a cocoa bean or a pan or anywhere else—turns brown when heated is to do with the presence of carbon. Sugars are carbohydrates, which is to say that they are made of carbon ("carbo-"), hydrogen ("hydr-"), and oxygen ("-ate") atoms. When heated, these long molecules disintegrate into smaller units, some of which are so small that they evaporate (which accounts for the lovely smell). On the whole, it is the carbon-rich molecules that are larger, so these get left behind, and within these there is a structure called a carbon–carbon double bond. This chemical structure absorbs light. In small amounts it gives the caramelizing sugar a yellow-brown color. Further roasting will turn some of the sugar into pure carbon (double bonds all round), which creates a burnt flavor and a dark-brown color. Complete roasting results in charcoal: all of the sugar has become carbon, which is black.

Another type of reaction, which occurs at a higher temperature, also contributes to the color and flavor of the cocoa: the Maillard reaction. This is when a sugar reacts with a protein. If carbohydrates are the fuel of the cellular world, proteins are the workhorses: the structural molecules that build cells and all their internal workings. Seeds (in the form of nuts or beans) must contain all of the proteins needed to get the cellular machinery of a plant up and running, so there is plenty of protein in the cocoa beans. When subjected to temperatures of 160°C and above, these proteins and carbohydrates start to undergo Maillard reactions, reacting with the acids and esters (produced by the earlier fermentation process) and resulting in a huge range of smaller flavor molecules. It is no exaggeration to say that without the Maillard reaction the world would be a much less delicious place: it is the Maillard reaction that is responsible for the flavor of bread crust, roasted vegetables, and many other roasted, savory flavors. In this case, the Maillard reaction is responsible for the nutty, meaty flavors of chocolate, while also reducing some of the astringency and bitterness.

Grind up the fermented and roasted cocoa nuts and add them to hot water and you have the original hot *chocolatl* made by the Mesoamericans. The Olmecs and then the Mayans, who first cultivated chocolate, drank it this way, and it was revered as a ceremonial drink and an aphrodisiac for hundreds of years. The cocoa nuts were even used as currency. When European explorers got hold of the drink in the seventeenth century, they exported it to coffee houses, where it competed with tea and coffee to be the beverage of choice of Europeans — and lost. What no one had really mentioned was that *chocolatl* means "bitter water," and even though it was sweetened with the new cheap sugar flooding in from the slave-run plantations of Africa and South America, it was also a gritty, oily, and heavy drink, because 50 percent of the cocoa bean is cocoa fat. This is how it remained for another two hundred years: an exotic drink, notable but not terribly popular.

With the invention of a few industrial processes, though, chocolate's fortunes suddenly changed. The first was the screw press, invented by a Dutch chocolate company called Van Houten in 1828. Crushing the fermented roasted beans with this press forced the cocoa butter to flow out and allowed Van Houten to separate it from the remaining cocoa solids. Now free of much of its fat, the cocoa could be ground down into a much finer powder and so lost its grittiness, becoming smooth, sleek, and velvety. It was in this form that cocoa now became popular — and survives to this day — as drinking chocolate.

Then came a moment of counterintuitive genius: having removed and purified the cocoa fat, and having pulverized the cocoa powder separately, why not mix them back together again, adding in some sugar, to create an ideal cocoa bean — the kind of bean you

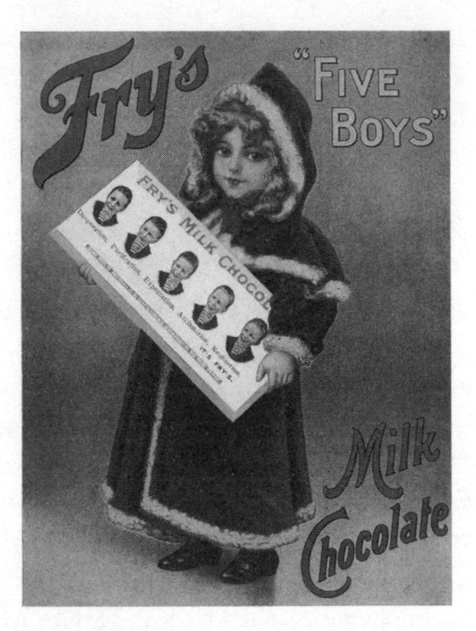

An advert for Fry's chocolate, 1902.

would want to pick from a tree, the kind of bean with exactly the right combination of sugar, chocolate flavor, and fat that would exist in a Willy Wonka world?

There were many chocolatiers in Belgium, Holland, and Switzerland who experimented with this approach, but it was an English firm called Fry and Sons that became famous for producing such nodules of "eating chocolate" and in doing so created the first chocolate bars. As the purified cocoa butter melted in the mouth, it released the cocoa powder, producing instant hot chocolate—a sensation that was completely unique. Because the cocoa fat content could be controlled separately from the cocoa powder and the sugar, it was now possible to design different types of sensations in the mouth to suit different tastes. And in a time before refrigerators, the cocoa butter's antioxidant properties meant that chocolate made in this way had a long enough shelf life to become a commercial product. The chocolate industry was born.

For some, even with the addition of 30 percent sugar, this form of chocolate was still too bitter, and so another ingredient was added, one that profoundly affected its taste: milk. This reduces the chocolate's astringency quite considerably, giving the cocoa an altogether milder—and the resulting chocolate an altogether sweeter—flavor. The Swiss were the first to do this in the nineteenth century, adding the plentiful milk powder produced by the fledgling Nestlé company, which itself was transforming milk from a local fresh product with a short life into a transportable commodity with a long one. The merging of the two commercial products, both with a long shelf life, was an enormous hit.

These days the type of milk added to chocolate varies widely throughout the world, and this is the main reason that milk chocolate tastes different from country to country. In the USA the milk used has had some of its fat removed by enzymes, giving the chocolate a cheesy, almost rancid flavor. In the UK sugar is added to liquid milk, and it is this solution, reduced to a concentrate, that is added to the chocolate, creating a milder caramel flavor. In Europe powdered milk is still used, giving the chocolate a fresh dairy flavor with a powdery texture. These different tastes do not travel well. Despite globalization, the preferred taste of milk chocolate, once acquired, remains surprisingly regional.

One thing that all milk chocolate has in common, though, is that almost all of the milk's water content has been removed before it is added. This is because chocolate powder is hydrophilic (water loving): given a chance it will absorb water, but in doing so it will eject its fat coating (water and fat will not dissolve in one another), in the process decomposing into a lumpy liquid much like the Mayan *chocolatl*. Anyone who has ever tried to add water to melted chocolate to create a sauce will have experienced this problem.

There are plenty of people, including myself, who are addicted to eating chocolate, and the reason may not just be its taste. It also contains psychoactive ingredients. The most familiar one is caffeine, which is present in small proportions in the cocoa bean, and so ends up in the chocolate via the cocoa powder. The other psychoactive ingredient is theobromine, which is a stimulant and antioxidant, like caffeine, but is also highly toxic to dogs. Many dogs die every year from eating chocolate, mainly around Easter and Christmas. Theobromine's effect on humans appears to be much milder, and the stimulant levels in chocolate are small when compared to coffee and tea, so even if you eat a dozen chocolate bars every day, it is only equivalent to drinking one or two cups of strong coffee. Chocolate also contains cannabinoids, which are the chemicals responsible for the high experienced from smoking dope. But again the percentages are tiny, and when blind taste studies were carried out to analyze chocolate cravings, researchers found little evidence that any of these chemicals were linked to feelings of craving.

This leaves another possibility to explain chocolate addiction. Rather than its being a chemical effect, it may be that the sensory experience of eating chocolate is itself addictive. Chocolate is like no other food. When chocolate melts in the mouth it suddenly releases a wild and complex, sweet and bitter cocktail of flavors within a warm rich liquid. It is not just a flavor but an entire oral experience. It is soothing and comforting, but it's also exciting and—not to put too fine a point on it—seems to satisfy more than a physical hunger.

Some say that eating chocolate is better than kissing, and scientists have dutifully tested this hypothesis by carrying out a set of experiments. In 2007, a team led by Dr. David Lewis recruited pairs of passionate lovers, whose brain activity and heart rate were monitored first while they kissed each other and then while they ate chocolate (separately). The researchers found that although kissing set the heart pounding, the effect did not last as long as when the participants ate chocolate. The study also showed that when the chocolate started melting, all regions of the brain received a boost far more intense and longer lasting than the brain activity measured while kissing.

Although this is just a single study, it does give credibility to the hypothesis that for many the sensory experience of eating chocolate is better than kissing. This association of chocolate with extreme sensory pleasure has been energetically promoted by chocolate manufacturers, most notably, perhaps, in the long-running television adverts for Cadbury's Flake chocolate bar.

The first Flake advert I ever saw featured a woman having a terrific time in the bath. I was young at the time, and baths did not fill me with the kind of delight that this woman was experiencing. For me baths were functional and usually cold, since my three older brothers had been in the bath before me. This was the 1970s, energy was expensive, and hot water was in short supply in our house. Baths were only jolly when I was allowed to bring my boats in with me. The woman on the TV didn't have toys of any kind but instead was equipped merely with a Flake chocolate bar. Every time she put some of it into her mouth, waves of contentment seemed to possess her, which would give her what seemed like the purest kind of pleasure. I realized that I had never experienced this kind of sensation at all, let alone in a bath. The advert had a strong impact on me and my brothers, and we tried to get our mum to allow us to eat chocolate in our cold baths, but without any luck. Instead she banned us from watching the adverts, an unenforceable directive because we didn't have a TV, and only saw the "lady of the Flake" when we stayed at our friends' houses. It was only much later that I realized it wasn't chocolate in the bath that she was trying to protect us from.

These adverts, which began in the late 1950s and continue to this day, always feature a woman relaxing on her own while indulging in the secret pleasure of eating a Flake. The shape and size of this rod-like chocolate, and the suggestive manner in which the woman indulged in it, were enough to send waves of outrage and alarm through the viewing public, despite the fact that the adverts never showed any nudity (merely implying it). It was, after all, an exercise wholly in suggestion. Indeed, a search on YouTube, where the original adverts have been uploaded, shows that the early versions were far more suggestive than recent ones. But while the call to censor these adverts was successful, their essential message has remained, and it does seem to resonate with the public, perhaps even pointing to a genuine truth about chocolate: for many, it is better than sex.

The actress Donna Evans in a 1960s Flake advert.

In a list of the countries with the highest consumption of chocolate, Switzerland comes top, followed by Austria, Ireland, Germany, and Norway. In fact, sixteen of the twenty countries with the highest chocolate consumption are Northern European. (In America, chocolate is more popular as a flavor than as a bar, with more than half the population saying they preferred chocolate drinks, cakes, and biscuits than any other flavor.) Given the reputation of chocolate as a substitute for sex, it is tempting to draw all sorts of cultural conclusions from this correlation. But there is another possible explanation for the high chocolate consumption in these countries, which is also associated with temperature.

In order to transform from a solid to a liquid easily within the mouth, chocolate requires a fairly cool ambient temperature. In a climate that is too warm, chocolate will either melt on the shelf or need to be put in the fridge, which defeats the purpose entirely—cold chocolate gets swallowed before it's had a chance to melt. (This problem may explain, perhaps, why the Mesoamericans, who first invented chocolate in the tropics, never created a solid bar but consumed it only as a drink.) Moreover, if solid chocolate is exposed to temperatures above 20°C, as a result perhaps of being left in the sun or in a hot car, it undergoes fundamental changes of structure. The changes can be spotted immediately because they result in "bloom": fat and sugars migrate to the surface of the chocolate and form a whitish crystalline powder, often with a river mark pattern.

As well as pure pleasure, chocolate's high sugar content and the perceived stimulating effects of the caffeine and theobromine have carved out another role for chocolate, encapsulated by the slogan "A Mars a day helps you work, rest, and play"; or its French equivalent: *"Un coup de barre? Mars et ça repart!"* ("Feeling beat? A Mars and you're off again!"); or its German one: *"Nimm Mars, gib Gas"* ("Take Mars, step on the gas"). With an average chocolate bar containing more than 50 percent sugar and 30 percent fat, it clearly offers a concentrated source of energy and an instant lift. For these same reasons, though, the healthiness of chocolate-rich diets has been called into question.

Cocoa butter is a saturated fat, a class of fat associated with an increased risk of heart disease. Further investigation into how the body digests this fat has shown, however, that it tends to convert this fat into an unsaturated fat, which is thought to be benign. Meanwhile, the cocoa particles contain an enormous range of antioxidants, and no one really knows what they get up to in the body. However, controlled studies by Harvard University have shown that the regular consumption of a small amount of dark chocolate leads to an increased life expectancy (as compared to the consumption of no chocolate at all). No one knows why, and further studies are ongoing. Of course, if the craving for chocolate becomes too much, any benefits will be offset by weight gain. At the moment, the jury is still out, but leaving aside over-consumption, chocolate is no longer seen as damaging to our health and perhaps even as beneficial.

For all these reasons, although we are a long way from doctors prescribing chocolate or children being given it as part of a school diet, chocolate is an integral item in many countries' standard military rations: it provides a sugar boost for energy, caffeine and theobromine for brain stimulation, and fats to replenish those lost during extreme exercise, and it has a shelf life of several years. Finally, but most controversially, it also may stave off feelings of sexual frustration.

Personally, I eat chocolate obsessively every afternoon and every night. Whether this is down to the brainwashing I received watching Flake adverts, or a psychophysical addiction, or sexual repression due to a Northern European upbringing, I am not sure. I prefer to think it's because I truly appreciate that chocolate is one of our greatest engineering creations. It is certainly no less remarkable and technically sophisticated than concrete or steel. Through sheer ingenuity, we have found a way to turn an unpromising tropical rainforest nut that tastes revolting into a cold, dark, brittle solid designed for one purpose only: to melt in your mouth, flood your senses with warm, fragrant, bittersweet flavors, and ignite the pleasure centers of the brain. Despite our scientific understanding, words or formulae are not enough to describe it. It is as close as we get, I would say, to a material poem, as complex and beautiful as a sonnet. Which is why the Linnaean name for the stuff, *theobroma*, is so appropriate. It means "the food of the gods."

5 MARVELOUS

ONE DAY IN 1998 I walked into the lab just as one of the technicians was taking a piece of material out of the microscope. "I'm not sure if you're allowed to see this," he said, "so we'd better be on the safe side, otherwise I'm going to have to fill in a load of paperwork." He quickly covered up the material.

I was working for the US government at the time, in a nuclear weapons laboratory in the desert of New Mexico. Being a British citizen I had only the basic security clearance, and so there were areas in the laboratory complex I couldn't go. Most areas, in fact. But this was our lab, so the behavior of the technician was definitely odd. I knew better than to ask him more. This was the late 1990s, a time when Chinese espionage in US national laboratories was a very sensitive issue. The American scientist Wen Ho Lee had just been caught, indicted, and put in solitary confinement on a charge of stealing nuclear secrets for China.[1] I was regularly interviewed about the security issues surrounding my work, and my US colleagues were under increasing pressure to report any conversations that were out of the ordinary. For a British person like me with an inquiring mind and a fondness of joking around, asking unnecessary questions was a risky business. The material, though, was extraordinary, and although I only saw a small fragment of it for a mere second, I found it impossible to forget.

Our research team would regularly go for lunch together at the various cafeterias near the laboratory complex. This meant leaving the air-conditioned safety of our building and striking out into the bright desert to pick up our cars from the sun-drenched asphalt parking lot. On the way we passed high wire fences beyond which the orange sand, dotted with cacti, stretched out to an air force base. The place was unreal in so many ways, but made more so by its contrast with our mundane routines. Driving through the desert in a convoy of cars, which had been heated to boiling point by the unblinking sun, to the cafeterias serving Tex-Mex cuisine was one such routine. Day after day, we would chat about nothing, our conversation bleached by the heat. Day after day, the thought of

---

1 He was eventually only charged with improper handling of secure data, to which he pleaded guilty. The judge eventually apologized to him for the solitary confinement.

the mystery material would pop into my head and I would wonder what on earth it could be. The fact that I couldn't talk to anyone about it made it all the more difficult to forget.

I remembered it as being transparent, yet strangely opalescent—like a hologram of a jewel: a ghost material. I had definitely seen nothing like it before. Had it, I wildly speculated, been salvaged from some alien spacecraft? After a while I started to doubt I had seen it at all. Then I became paranoid that they might be trying to brainwash me into thinking it was all a figment of my imagination. "I have actually seen it," I kept saying to myself as I drove to and from the cafeteria day after day. I felt strangely proprietorial over it. Finally, I worried that it wasn't being treated right. It was that day that I decided I would have to leave.

I didn't see it again until a few years later. I was back in the UK, having taken a job as the head of the Material Research group at King's College London. One afternoon I was at home making a birthday card for my brother Dan, when they announced on the TV news that on January 2, 2004, the NASA mission to capture stardust had successfully engaged with the comet Wild 2. The news program then showed a picture of *my material*. Well, obviously not *my* material, but the material I desperately wanted to be mine. "So it was alien!" I said triumphantly to my empty flat, as I scrambled on to my computer to find out more. "They are harvesting it from space," I thought. Wrongly.

The material turned out to be a substance known as aerogel. I had got the wrong end of the stick from the news report: it was the aerogel that was being used to collect the stardust. I didn't really stop to think about this but plowed on, collecting information about aerogels and their history. Aerogels were not of alien origin, I found out, but they nevertheless had a very strange back-story: they were invented in the 1930s by a man called Samuel Kistler, an American farmer turned chemist, who conjured them into existence solely to satisfy his curiosity about jelly. Jelly?

What was jelly? he asked. He knew that it wasn't a liquid, but it wasn't really a solid either. It was, he decided, a liquid trapped in a solid prison, but one in which the prison bars were like an invisibly thin mesh. In the case of edible jelly, the mesh is made from long molecules of gelatin, which is derived from the protein, collagen, that makes up most connective tissues, such as tendons, skin, and cartilage. When added to water, these gelatin molecules unravel and connect with one another to form a mesh that traps the liquid within it and prevents it from flowing. Thus jelly is basically like a water balloon, but instead of being an outer skin that holds the water within, it inhabits the water from the inside.

The water is held inside the mesh by a force known as surface tension—the same force that makes water feel wet and form drops, and causes it to stick to things. The surface tension forces inside the mesh are strong enough for the water to be unable to escape the jelly, but weak enough for it to slosh around—which is why jelly wobbles. It's also why jelly feels so amazing when you eat it: it's almost 100 percent water, and with a melting point of 35°C the internal gelatin network promptly melts, freeing the water to burst in your mouth.

The simple explanation—a liquid trapped by a solid internal mesh—was not enough for Samuel Kistler. He wanted to know whether the invisible gelatin mesh within a jelly was all of a piece. In other words, was it a coherent, independent internal skeleton, such that if you could find a way to remove all of the liquid from it, the mesh could stand on its own?

To answer the question he conducted a series of experiments, the results of which he published in a letter to the scientific journal *Nature* in 1931 (no. 3211, vol. 127, p. 741). The letter is entitled "Coherent Expanded Aerogels and Jellies," and here is how he introduced the report:

"The continuity of the liquid permeating jellies is demonstrated by diffusion, syneresis, and ultra-filtration, and the fact that the liquid may be replaced by other liquids of very diverse character indicates clearly that the gel structure may be independent of the liquid in which it is bathed."

What he is saying in this opening paragraph is that various experiments have shown that the liquid in a jelly is connected throughout, rather than being compartmentalized, and can be replaced by other liquids. This demonstrates, in his opinion, that the solid internal skeleton may indeed be independent of the liquid in the jelly. And in using the word "gel," as a more general word for jelly, he is saying that this is true of a whole range of jelly-like materials that span the gap between being truly solid and truly liquid, from hair gel, to solid chicken stock, to setting cement (where the internal mesh is formed by calcium silicate fibrils).

He goes on to point out that no one had yet managed to separate the liquid of a jelly from its internal skeleton: "Hitherto the attempt to remove the liquid by evaporation has resulted in shrinkage so great that the effect upon the structure may be profound." In other words, those in the past who have tried to remove the liquid by evaporation have found that the internal skeleton simply collapses. He then goes on to say triumphantly that he and his collaborators have found a way to do it:

"Mr. Charles Learned and I, with the kindly assistance and advice of Prof. J. W. McBain, undertook to test the hypothesis that the liquid in a jelly can be replaced by a gas with little or no shrinkage. Our efforts have met with complete success."

Their cunning idea was to replace the liquid with a gas while it was still inside the jelly, and so use the pressure of the gas to keep the skeleton from collapsing. First, though, they found a way to replace the water in the jelly with a liquid solvent (they used alcohol), which would be easier to manipulate. The danger of using a liquid solvent was that it too would evaporate, but they found a way to stop it:

"Mere evaporation would inevitably cause shrinkage. However, the jelly is placed in a closed autoclave with an excess of liquid and the temperature is raised above the critical temperature of the liquid, while the pressure is maintained at all times at or above the vapor pressure, so that no evaporation of liquid can occur and consequently no contraction of the gel can be brought about by capillary forces at its surface."

An autoclave is simply a pressure tank that can be heated. By increasing the pressure in the autoclave, the liquid inside the jelly is prevented from evaporating, even when the temperature is increased beyond its boiling point. The capillary forces he talks about, meanwhile, are caused by the surface tension of the liquid. Kistler speculates that when the liquid is gradually removed through evaporation, these same forces that hold the jelly together are responsible for tearing it apart. But when he raises the temperature of the whole jelly above the "critical temperature"—the point at which there is no difference between a gas and a liquid because both have the same density and structure—the whole liquid becomes a gas without going through the destructive process of evaporation. He says:

"When the critical temperature is passed, the liquid has been converted directly into a permanent gas without discontinuity. The jelly has had no way of 'knowing' that the liquid within its meshes has become a gas."

This is a stroke of genius: under the pressure from the autoclave, the newly created gas cannot escape from the jelly and so the internal skeleton stays intact.

"All that remains is to allow the gas to escape, and there is left behind a coherent aerogel of unchanged volume."

Only now does he let the gas escape slowly, leaving the internal skeleton of the jelly completely intact and mechanically sound, thus proving his hypothesis. It must have been a very satisfying moment. But he didn't stop there. These internal skeletons of jelly were incredibly light, fragile things, comprised mostly of air. They were, in fact, foams. Perhaps he could make them stronger, he thought, by making a jelly not out of gelatin but out of something

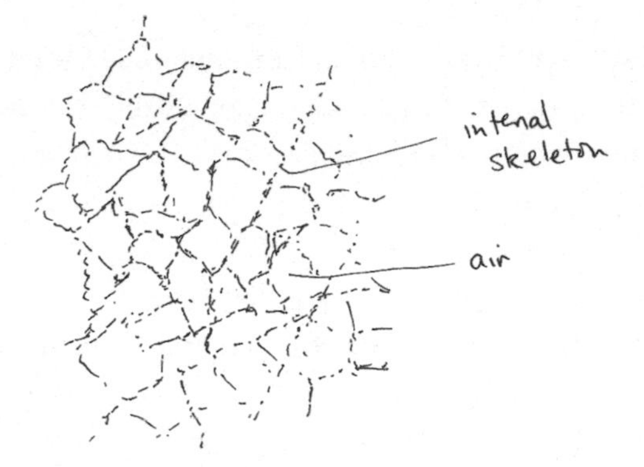

Internal skeleton of a jelly.

more rigid. So it was that he engineered a jelly in which the internal skeleton was made of the mineral silicon dioxide: the main constituent of glass. Using exactly the same process described above, he then created from this jelly a "silica aerogel": the lightest solid in the world. This was the material I had seen for a split second all those years ago in a laboratory in the desert.

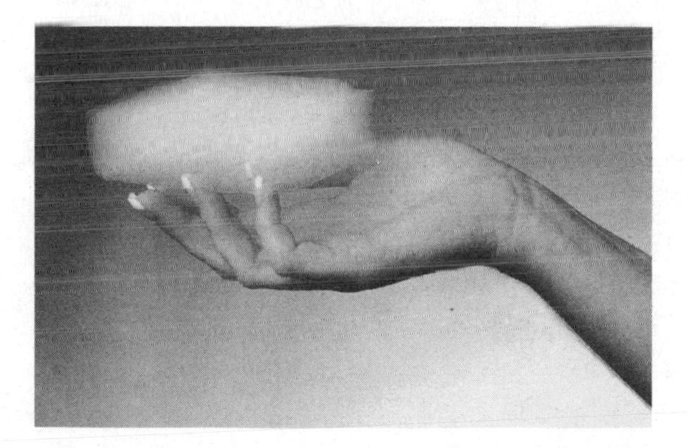

Silica aerogel, the lightest solid in the world, which is 99.8 percent air.

Not content with this achievement, Kistler went on to make other aerogels, and he lists them in the paper:

"So far, we have prepared silica, alumina, nickel tartarate, stannic oxide, tungstic oxide, gelatine, agar, nitrocellulose, cellulose, and egg albumin aerogels and see no reason why this list may not be extended indefinitely."

Note that despite his triumph with silica aerogel he couldn't resist making an aerogel from egg albumin—that's egg white. So while the rest of the world were using egg whites to cook light fluffy omelets and bake cakes, Kistler did a different type of cooking using an autoclave to create egg aerogel: the lightest meringue in the world.

Silica aerogel looks extremely odd. Put it against a light background and it disappears almost entirely. In this sense, it is harder to see—more invisible, even—than normal glass despite being less transparent. When light passes through glass, its path is distorted slightly—it is refracted—and the degree of distortion is known as glass's refractive index. In the case of aerogel, because there is simply less of the stuff, light's path is hardly distorted at all. For this same reason, there is no hint of reflection on its surfaces, and because of its ultra-low density it appears to have no distinct edges, to not be fully solid at all. Which of course it isn't. The internal skeleton of a jelly has a structure not unlike that of bubble bath foam, with one main difference, which is that all of the holes link up. Silica aerogel is so full of holes that it is typically 99.8 percent air and has a density only three times greater than air, which means that it has practically no weight at all.

At the same time, when placed against a dark background silica aerogel is undoubtedly blue. And yet, since it is made from clear glass, it ought to have no color at all. For many years, scientists wondered why this might be. The answer, when it came, was rather satisfyingly odd.

When light from the sun enters the Earth's atmosphere, it hits all sorts of molecules (mostly nitrogen and oxygen molecules) on its way to Earth and bounces off them like a pinball. This is called scattering, which means that on a clear day, if you look at any part of the sky, the light you see has been bouncing around the atmosphere before coming into your eye. If all light was scattered equally, the sky would look white. But it doesn't. The reason is that the shorter wavelengths of light are more likely to be scattered than the longer ones, which means that blues get bounced around the sky more than reds and yellows. So instead of seeing a white sky when we look up, we see a blue one.

This Raleigh scattering, as it is called, is very slight indeed, so you need an enormous volume of gas molecules to see it: the sky works but a room full of air doesn't. Put another way, any one bit of the sky doesn't look blue but the whole atmosphere does. But if

Silica aerogel protecting a flower from the high temperatures of a Bunsen burner.

a small amount of air happens to be encapsulated in a transparent material that happens to contain billions and billions of tiny internal surfaces, then there will be sufficient Raleigh scattering off these surfaces to change the color of any light that passes through it. Silica aerogel has exactly this structure, and this is where its blue hue comes from. So when you hold a piece of aerogel in your hand, it is, in a very real way, like holding a piece of sky.

Aerogel foams have other interesting properties, the most remarkable of which is their thermal insulation—their ability to act as a barrier against heat. They are so good at this that you can put the flame of a Bunsen burner on one side of a piece of aerogel and a flower on the other and still have a flower to sniff a few minutes later.

Double glazing works by providing a gap between two glass panes which makes it hard for heat to conduct between them. Imagine that the atoms in glass are arranged like the audience in a rock concert, all packed together and dancing around. As the music gets louder and the audience dances more energetically, people knock into each other more. The same happens in the glass: as the material heats up, the atoms jiggle about more. The definition of the temperature of a material is, in fact, the degree to which the atoms in it are jiggling around. In the case of double glazing, though, there is a gap between the two glass planes, which means that the jiggling glass atoms in one pane find it hard to pass on their energy to those in the other. Of course, this works both ways: the same double glazing can be used to keep heat inside a building in the Arctic and to keep it outside a building in Dubai.

Double-glazed windows work well enough, but they still leak a lot of heat—as anyone who lives in a hot or cold country knows by looking at their energy bill. Could we do better? Well, there is, of course, triple glazing and quadruple glazing, which work by introducing a new layer of glass and so a new barrier to the heat transfer. But glass is dense, so these windows get heavier and bulkier and less transparent the more layers there are. Enter aerogel. Because it is a foam, it has within it the equivalent of a billion billion layers of glass and air between one side of the material and the other. This is what makes it such a superb thermal insulator. Having discovered this and other remarkable properties, Kistler reported them in the final sentence of his paper as follows:

"Apart from the scientific significance of these observations, the new physical properties developed in the materials are of unusual interest."

Unusual interest, indeed. He had discovered the best insulator in the world.

The scientific community applauded briefly, but then promptly forgot all about aerogels. It was the 1930s and they had other fish to fry; it was hard to know what would shape the future and what would be forgotten. In 1931, the year Kistler reported his invention of aerogels, the physicist Ernst Ruska created the first electron microscope. In the same edition of *Nature* in which Kistler published his findings, the Nobel prizewinning materials scientist William Bragg reported his findings on the electron diffraction within crystals. These scientists paved the way for a new understanding of the inner structure of materials by developing the tools with which to see and visualize them. It was the first time that a new microscope had been invented since the optical microscope of the sixteenth century and a whole new microscopic world was opening up. Soon, materials scientists were peering into metals, plastics, ceramics, and biological cells, and starting to understand how they worked from an atomic and molecular perspective. It was an exciting time: the world of materials was exploding and materials scientists would soon deliver nylon, aluminum alloys, silicon chips, fiberglass, and many other revolutionary materials. Somehow in all the excitement aerogels got lost and everyone forgot about them.

Everyone except one man, Kistler himself. He decided that the beauty and thermal insulation properties of these jelly skeletons were so extraordinary that they should and must have a future. Although silica aerogel is as fragile and brittle as glass, for its weight (which is minuscule), it has good strength, certainly enough to make it industrially useful. So he patented it and sold the license to manufacture it to a chemical company called Monsanto Corporation. By 1948 it was making a product called Santogel, which was a powdered form of silica aerogel.

Santogel seemed to have a bright future as the best thermal insulator in the world, but alas the time was not right for it. Energy was getting cheaper and cheaper, not more expensive, and there was no awareness of the problem of global warming. An expensive thermal insulator like aerogel just didn't make economic sense.

Having failed to find a market in thermal insulators, Monsanto rather bizarrely found applications for it in various inks and paints, its role being to flatten them optically by scattering light, creating a matte finish. Aerogel finally ended up being used ignominiously, as a thickening agent in screw-worm salves for sheep and in the jelly used to create napalm for bombs. In the 1960s and 1970s, cheaper alternatives usurped aerogel even from this rather limited repertoire of applications, and finally Monsanto gave up making it altogether. Kistler died in 1975 having never seen his most wonderful material find a place in the world.

The revival of aerogels came not as a result of any commercial application but because their unique properties attracted the attention of some particle physicists at CERN studying something called Cherenkov radiation. This is the radiation given off by a subatomic particle when it travels through a material faster than light can travel through it. Detecting and analyzing this radiation gives clues to the nature of the particle and so provides a very exotic means of identifying which of the many invisible particles the scientists are dealing with. Aerogel is extremely useful for this purpose—providing a material through which the particle can travel—as it is, effectively, a solid version of a gas, and it continues to be used for this today, helping physicists unravel the mysteries of the subatomic world. Once aerogels found their way into physicists' labs, with their sophisticated equipment, esoteric aims, and big budgets, the material's reputation started to grow again.

At that time in the early 1980s, aerogels were so expensive to make that they could only live in labs where money was no object. CERN was one such lab, but soon NASA followed. The first applications of silica aerogels in space exploration were to insulate equipment from extreme temperatures. Aerogels are particularly suitable for this application because not only are they the best insulators in the world, but they are also extremely light, and when you're launching spacecraft out of the gravitational pull of the Earth, reducing weight matters rather a lot. Aerogel was used first in 1997 on the Mars Pathfinder mission and has been used as an insulator on spacecraft ever since. But once the scientists at NASA found that aerogel could cope with space travel, they realized that the material had another possible use.

If you look up into the sky on a clear night you might see a shooting star, which appears as a bright trail of light crossing the sky. For a long time it has been known that these are meteors which enter the Earth's atmosphere at high speeds and burn brightly as they heat up. It is thought that most of these are space dust, which is leftover material from the creation of the solar system 4.5 billion years ago, along with comets and other asteroids. Determining exactly what materials these heavenly bodies are made from has been of interest for many years, since this information could help us understand how the solar system was formed and may also account for the chemical composition of the Earth.

Analyzing the material composition of meteorites has given us some tantalizing clues, but the problem with these specimens is that they have all been heated to extremely high temperatures by their passage through the atmosphere. Wouldn't it be nice, the people at NASA thought, if they could capture some of these objects out in space and bring them back to Earth in a pristine state?

The first problem with this idea is that objects in space tend to be traveling rather fast. Space dust is often going at speeds of fifty kilometers per second, which equates to eighteen thousand kilometers per hour, a lot faster than a bullet. Catching an object like that is not easy. As with stopping a bullet with, say, your body, either the force of the bullet exceeds the rupture pressure of your skin, meaning it goes through you, or you employ a bulletproof vest made of a high-rupture-strength material, such as Kevlar, which results in a compressed and deformed bullet. Either way, it's a risky business. However, in principle, it is quite possible—just as when catching a cricket ball or baseball with "soft" hands, the trick is to spread and dissipate the ball's energy rather than bracing yourself for a single, high-pressure impact. What NASA needed, then, was a way to slow the dust down from eighteen thousand kilometers per hour to zero without damaging the dust or the spacecraft—ideally a material with a very low density, so that the dust particles would be slowed gently without being damaged; ideally one that could do so within the space of a few millimeters; and ideally one that would be transparent, so that scientists could find the tiny specks of dust once they were buried in it.

That such a material existed was a minor miracle. That NASA had already used it in space flights was extraordinary. It was, of course, silica aerogel. The mechanism by which aerogel pulls off this feat is the same as the one used to protect stunt actors in movies when they fall off tall buildings: a mountain of cardboard boxes, each box absorbing some of the energy of the impact as it collapses beneath the actor's weight, and the more boxes, the better. In the same way, each foam wall within aerogel absorbs a tiny amount of energy when it is struck by the dust particle, but since there are billions of them per cubic centimeter, there are enough of them to bring it to a halt relatively unharmed.

NASA built an entire space mission around the ability of aerogel to gently collect stardust. On February 7, 1999, the Stardust spacecraft was launched, containing all of the equipment necessary to take a trip through the solar system, while also being programmed to fly past a comet called Wild 2. The idea was that it would collect interstellar dust from deep space as well as the dust being ejected from a comet, allowing NASA to study the material composition of both. In order to do this, they developed a tool that resembled a giant tennis racket, but instead of holes between the strings there was aerogel.

During the summer and autumn of 2002, while in deep space, millions of kilometers from any planet, the Stardust spacecraft opened a hatch and poked out its giant tennis racket fitted with aerogel. It had no opponent in this game of interstellar tennis and the balls it was looking for were microscopically small: the remains of other stars long gone, the leftover ingredients of our own solar system still flying around. The Stardust spacecraft couldn't hang around in deep space too long because it had an appointment to keep with the comet Wild 2, now hurtling from the outer reaches of the solar system and approaching the center, which it does every 6.5 years. Having withdrawn its aerogel tennis racket, the spacecraft sped off for its meeting. It took just over a year to get to the right position, but on January 2, 2004, the spacecraft found itself on a collision course with the comet, which was five kilometers in diameter and speeding off around the sun. Once it had maneuvered itself into the slipstream of the comet, 237 kilometers behind it, the spacecraft opened its hatch and once again poked out its aerogel tennis racket, this time using the B-side, and started to collect, for the first time in human history, virgin comet dust.

Having collected the comet dust, the Stardust spacecraft returned to Earth, arriving back two years later. As it approached the Earth it veered away, jettisoning a small capsule, which fell under Earth's gravity, entering the atmosphere at a speed of 12.9 kilometers per second, the fastest re-entry speed ever recorded, and so becoming for a while a shooting star itself. After fifteen seconds of free-fall, and having reached red-hot temperatures, the capsule deployed a drogue parachute to slow down the rate of descent. A few minutes later, at a height of ten thousand feet above the Utah desert, the capsule jettisoned the drogue chute and deployed the main parachute. At this point the recovery crews on the ground had a good idea of where the capsule was going to land and headed out into the desert to welcome it back from its seven-year, four-billion-kilometer round trip. The capsule hit the sand of the Utah desert at 10:12 GMT on Sunday, January 15, 2006.

"We feel like parents awaiting the return of a child who left us young and innocent, who now returns holding answers to the most profound questions of our solar system," said the project manager, Tom Duxbury, of NASA's Jet Propulsion Laboratory in Pasadena, California.

However, until they opened the capsule and started examining the aerogel samples, scientists had no idea whether they held any answers to anything. Perhaps the space dust would have passed straight through the aerogel. Or perhaps the violence and deceleration of re-entry would have disintegrated the aerogel into meaningless powder. Or perhaps there would be no dust at all.

They need not have worried. Once they got the capsule back to the NASA laboratories and opened it up, they found that the aerogel was fully intact and almost completely perfect. There were minuscule puncture marks in the surface and it was these that were subsequently shown to be the entry points for the space dust. Aerogel had done the job that no other material could do: it had brought back pristine samples of dust from a comet formed before the Earth even existed.

Since the return of the aerogel capsule, it has taken NASA's scientists many years to find the tiny pieces of dust embedded within the aerogel, and the work continues to this day. The dust they are

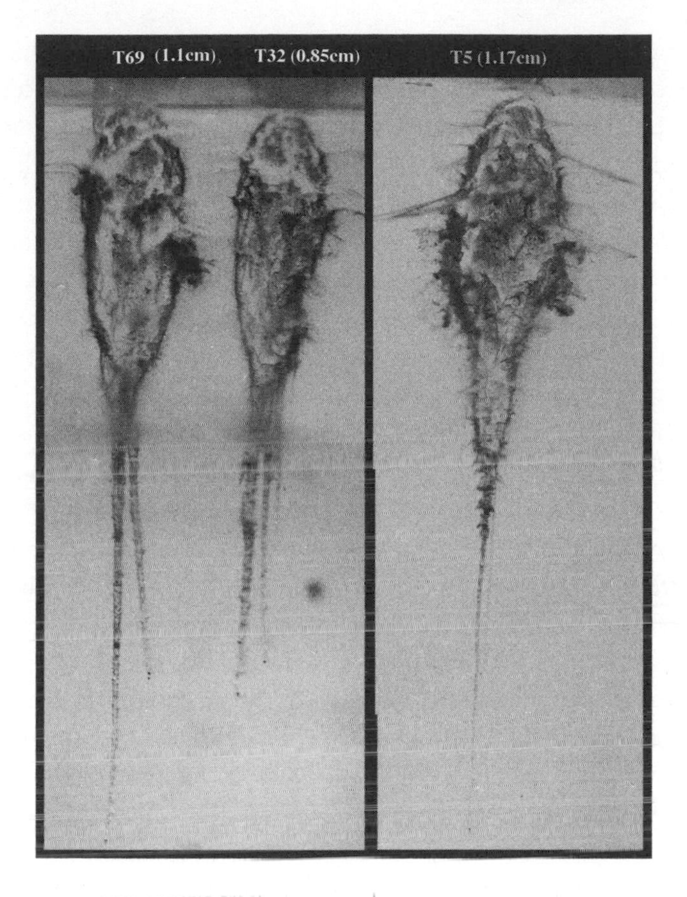

Microscopic comet particle tracks in aerogel.

looking for is invisible to the naked eye, and so it must be found by microscopic examination of the samples, which has taken years. The project is so massive that NASA has enlisted the public to help with the search. The scheme Stardust@Home trains volunteers to use their home computers to look through thousands of microscopic images of the aerogel samples and try to spot the signs that a piece of space dust is present.

The work so far has thrown up a number of interesting results, the most surprising of which is that most of the dust from the comet Wild 2 shows the presence of aluminum-rich melt droplets. It's very hard to understand how these compounds could have formed in a comet that had only ever experienced the icy conditions of space, since they require temperatures of more than 1200°C to do so. Since comets are thought to be frozen rocks that date back to the birth of the solar system, this has come as a bit of a surprise, to say the least. The results seem to indicate that the standard model of comet formation is wrong, or there is a lot more we don't understand about how our solar system formed.

Meanwhile, having completed its mission, the Stardust spacecraft has now run out of on-board fuel. On March 24, 2011, when it was 312 million kilometers away from Earth, it responded to a final command from NASA to shut down communications. It acknowledged this command, and said its final goodbye. It is currently traveling off into deep space, a kind of man-made comet.

Now that the Stardust mission is over, will this be the fate of aerogel too, to end in obscurity? It is all too possible. Although aerogels are the best insulators we have, they are very expensive and it is not clear that even now we care about energy conservation enough to value aerogels economically. There are several companies selling aerogel for such thermal insulation applications, but at the moment the main ones are for extreme environments such as drilling operations.

It's possible that, because of environmental considerations, our energy costs will get higher and higher. In a sufficiently high-cost energy future, it is conceivable that the monolithic double glazing we are all used to may be replaced with a much more sophisticated glass material based on aerogel technology. Research on developing new aerogels has been taking place at an increasingly rapid pace. There are now a number of aerogel technologies that result in a material that is not rigid and brittle, as silica aerogels are, but flexible and bendy. These so-called x-aerogels are made flexible by a neat piece of chemistry that detaches the rigid foam walls of an aerogel from one another and inserts between them polymer molecules that act like hinges within the material. These x-aerogels can be made into flexible materials such as textiles and could be used to make the warmest but lightest blankets in the world, potentially replacing duvets, sleeping bags, and the like. Because they are so light they would also be perfect for outdoor clothes and boots designed for extreme environments. They could even replace the foam soles in sports shoes that make that type of footwear so springy. Recently, a family of carbon aerogels have been created which conduct electricity, as well as super-absorbent aerogels that can suck up toxic waste and gases.

So aerogels may yet be part of our everyday lives, the answer perhaps to living in a more extreme and volatile climate. But although as a materials scientist it's good to know that we are likely to have the right materials to offer the world in the event that global warming is not averted, this is not the kind of future I want for my children. In a world where we have industrialized so many materials, including those we used to hold sacred, such as gold and diamond, I like to think there may again be a place for a material valued solely for its beauty and significance. Most people will never hold a piece of aerogel in their hand, but those who do never forget it. It is a unique experience. There is no weight to it that you can perceive, and its edges fade away so imperceptibly that it is impossible to see where the material stops and the air begins. Add to this its ghostly blue color and it really is like holding a piece of sky in your hand. Aerogels seem to have the ability to compel you to search your brain for some excuse to be involved with them. Like an enigmatic party guest, you just want to be near them, even if you can't think of anything to say. These materials deserve a different future, not of oblivion or embedment in a particle accelerator, but to be valued for themselves.

Aerogels were created out of pure curiosity, ingenuity, and wonder. In a world where we say we value such creativity, and give out medals to reward its success, it's odd that we still use gold, silver, and bronze to do so. For if ever there was a material that represented mankind's ability to look up to the sky and wonder who we are, if ever there was a material that represented our ability to turn a rocky planet into a bountiful and marvelous place, if ever there was a material that represented our ability to explore the vastness of the solar system while at the same time speaking of the fragility of human existence, if ever there was a blue-sky material—it is aerogel.

WHILE DOING MY PHD at the Materials Department of Oxford University, I regularly went to cinema matinees. Watching films in dark empty movie theaters soothed my brain in a way that no other form of relaxation could, especially on gray, rainy afternoons. One day, though, something strange happened. I had a blazing row with a stranger in the foyer. The film being screened that afternoon was *Butch Cassidy and the Sundance Kid,* a classic Western starring Paul Newman and Robert Redford. Before the movie started, I was in the queue to buy a bag of candy, when I heard the guy behind me bemoaning how sad it was that cinemas had lost their magic and become glorified shops selling overpriced confectionery. He was in the queue for this very same confectionery he professed to hate, I noted, but the inconsistency wasn't enough to override my British reserve. It's what he said next that really got to me. "Why are all the sweets sold in plastic bags?" he said. "In my day, they were paper. The sweets were stored in jars and sold by the weight in paper bags." I did a half-turn, clasping my bright orange shiny packet of chocolate, met his eye and, smiling, said, "But surely plastic has the most right of all materials to be in a cinema, especially in the context of this film." Maybe my tone was a bit condescending—it probably was. I was a PhD student and thought I knew quite a lot. But I was also keen to defend plastic, a material that is often misrepresented.

I had picked the wrong person, though. He tore into me with the full force of an authoritative film buff. What poppycock was I talking? What did I know about cinema at my age? He had lived and breathed it in its golden age. Cinema was about the stars of the silver screen, the flickering lights, the velvet seats, and the whir of the projector. He didn't listen to a word I said, and in truth he was probably right not to. Whatever the factual basis of my argument, I didn't have the language to convey it in a meaningful way. In the end, we both went into the cinema scarlet with rage and sat at opposite sides of a largely empty auditorium. I breathed a huge sigh of relief when the lights finally went down.

I have often thought of that humiliating episode, and wondered how I could have defended plastic more effectively. I have come to the conclusion that the only way he would have heard my argument is through the medium that he valued most: the visual language of cinema. So here it is, a screenplay to settle the argument that I lost about the relevance of plastic candy wrappers to the film *Butch Cassidy and the Sundance Kid.* At times, my rudimentary screenplay and its relevance to my cinema foyer argument are, perhaps, a little obscure, so each scene is followed by a few explanatory notes.

SCENE 1

Interior: Saloon bar, San Francisco, 1869.

It is early afternoon. Chairs and tables fill the room, half of which are occupied by men playing cards and drinking. There is a piano in the corner that no one is playing. The bright California sun filters through the blinds, which are broken and make a rattling noise when the wind blows. Cigar smoke lingers in the air.

The occupants of the saloon are a collection of rough-looking men, mostly out of work. Some are ex-miners who came out West during the California Gold Rush ten years earlier and then gravitated to the city, having failed to get rich. Others are veterans of the Civil War, who have wound up here as guns for hire. A few women keep them company.

In the corner there is a billiards table to accommodate the new craze for "pool," using fifteen colored billiard balls. **BILL** and his younger brother **ETHAN** are playing. **BILL** is a cowboy who came to the city on the run from killing a man in Ohio. He is a largely silent man with a toothless smile on account of having his teeth knocked out by a horse. He has persuaded his younger brother to travel out here with him, using the new railroad to make their journey.

ETHAN
(Over the pool table about to play a shot)
Blue ball, corner pocket.

BILL
(Leaning against the wall with cue in hand)
Yeah?

ETHAN

(Smashes ball into the corner pocket)

Yes siree! I like these new balls, I most definitely do.

BILL

Yeah?

ETHAN

Yeah, (adopts a fake posh accent) made for a man of leisure, don't you know — that's us, right, Bill? Men of leisure? Ha!

ETHAN goes on to pot a further two balls, keeping a running commentary in his fake posh accent and grinning at **BILL** in between each successful shot. **BILL** takes no notice, distracted by an argument at one of the card tables.

A **RED-FACED MAN,** a newcomer, has finally worked out he is being systematically cheated at cards and rises to his feet quickly but unsteadily as his chair tips over behind him and smacks violently on the floor. The rest of the table laugh. The **RED-FACED MAN** is so drunk that the thoughts going through his head are clearly visible on his face. He considers tipping up the card table and walking out, but then his hand finds his gun and he starts to wag it at the other players. The laughter stops, and after a few seconds everything in the saloon stops. Except **ETHAN,** who has his back to the room and has been lining up another seemingly impossible shot.

ETHAN

(Fake posh voice) Blue ball, end pocket.

In the silence he plays the shot, but something strange happens when the white ball hits the eight ball — there is a bright flash accompanied by a loud percussive crack. The eight ball misses the pocket, knocked off course by the mini-explosion.

THE RED-FACED MAN, whose befuddled attention had started to turn towards **ETHAN,** is so startled by the noise that he instinctively reacts by letting off one shot in the direction of the pool table before running out of the saloon.

ETHAN lies bleeding on the floor, as the white ball finally comes to a halt, still smoldering from its explosive collision with the eight ball.

NOTES FOR SCENE 1

The game of pool evolved from billiards, a fifteenth-century Northern European game that started in royal palaces and was essentially an indoor version of croquet. This is why the table surface was colored green, to simulate grass. One of the results of the Industrial Revolution was to make billiard tables much cheaper to produce. As in our day, it was found that the game could increase the income of bars and public taverns, and it started being adopted by the new urban poor.

During the nineteenth century the game got more technically sophisticated. First the cue sticks became tipped with leather and covered in chalk, to allow greater control of the ball by using spin. This technique was introduced to America by English sailors and is still referred to as putting "English" on the ball. In the 1840s the invention of vulcanized rubber by Charles Goodyear allowed for the introduction of "cushions" at the sides of the playing surface, which were soft and springy instead of wooden, ensuring that the balls would bounce off them in a predictable manner for the first time. From this point onwards billiard tables resembled those we know today. The move from a billiard game, which uses three or four balls, to the more fashionable game of pool, which uses fifteen balls, happened in 1870s America. Up until this point, though, the balls themselves were made of ivory and so were expensive.

Ivory has a unique combination of materials properties: it is hard enough to endure the thousands of high-speed collisions between the balls without denting or chipping; it is tough enough to not crack; it can be machined into the spherical shape of a ball; and like many organic materials it can be dyed into different colors. No other material at the time had this combination of properties, so when the popularity of pool exploded into a craze across

American saloons, there was a real chance that the price of ivory would rise so much in response to demand that the game would rapidly become unaffordable. Hence the trial in many saloons across the continent of balls made of new replacement materials, such as plastic, some of which behaved very strangely. Plastic was a new kind of material, as different from other materials as screenplays are to prose.

SCENE 2

Interior: A shack in downtown New York.

The shack is serving as a laboratory for **JOHN WESLEY HYATT,** a young man who works as a printer of newspapers, but in his spare time conducts chemical experiments. At the age of twenty-eight he already has one patent to his name, and is about to go into the history books as maker of the world's first usable plastic.

He is being visited by **GENERAL LEFFERTS,** a retired soldier and investor, who having already financially backed the young Thomas Edison is now interested in **HYATT. LEFFERTS,** a big formal man, is having to stoop to avoid hitting his head on the roof of the shack, which is full of glassware, wooden barrels, and a surprising amount of ivory. There is strong smell of solvents even though the windows are wide open.

HYATT

I got the idea for what I want to show you when I was trying to make synthetic billiard balls. (He gestures to the corner of the room, where there is a box of assorted billiard balls.)

LEFFERTS

Billiard balls? Why billiard balls?

HYATT

Currently they can only be made from ivory, but it's too expensive, and the game is so popular nowadays that the billiard table manufacturers are getting spooked that they will run out of it. So they put an advertisement in the

New York Times with a reward of ten thousand dollars for anyone who could invent a substitute material.

LEFFERTS
Ten thousand dollars! Shoot, there can't be that much money in the game.

HYATT, who is fiddling with his chemical apparatus, breaks off what he is doing to search for something and quickly locates it pinned to the wall. It is a yellowed newspaper clipping of the *New York Times* reward advertisement. He hands it to **LEFFERTS**.

HYATT
See for yourself.

LEFFERTS
(Reading the article while puffing away at a cigar) "Phelan & Collender, America's largest billiards supplier"—I've never heard of them . . . (Continues to read silently, mouthing some of the words, and then quotes a passage) "We offer a handsome fortune of ten thousand dollars to any inventor who can create a replacement material for ivory."
Well, well, can it be true?

HYATT
Oh, it's true, all right. I've been working on the problem for several years now and have supplied them with many prototypes. A few months ago they contacted me to say that they had sent several sets of my latest versions to saloons around the country on a trial basis.

LEFFERTS
So you succeeded?

HYATT
Well, yes . . . (Looks down as if not sure how to continue) But there's a problem . . . Er . . . let me show you how I made them and then you'll see. In fact, that's really why I brought you here, because you have to see this to believe it.

HYATT finishes off fiddling with several bits of his experimental kit and then takes out a large Dewar, a type of vacuum flask, from a locked cupboard and starts pouring from it a clear liquid into a beaker.

HYATT
This is the key to it all, and it was under my nose all the time!

LEFFERTS
What is it?

HYATT
It's a preparation of nitrocellulose in alcohol.

LEFFERTS
Nitrocellulose . . . I have heard of that . . . hmm . . . yes, but isn't that an explosive?!

LEFFERTS is suddenly red-faced, upset at his own naiveté in coming to visit this crazy scientist and putting his own life in danger. He fingers his cigar nervously—he has seen enough stupid accidents with explosives in the Civil War.

HYATT
(Not picking up on **LEFFERTS**'s concern) Oh, I think you mean nitroglycerin, yes, I suppose it is a little similar chemically, but this is nitrocellulose, which isn't technically explosive. Perhaps it's slightly explosive, highly flammable certainly. But I am very careful.

He turns to **LEFFERTS** at this point to smile, and then realizes that **LEFFERTS** is upset and so goes on to give further explanation by way of smoothing things over.

HYATT
Nitroglycerin is made by nitrating glycerol, which is an oily colorless liquid that results from soap manufacture. You just mix the glycerol with nitric acid. But, as you say, it is very unstable and is the key ingredient in dynamite. What I

have here, however, is nitrocellulose, which is made by mixing wood pulp with nitric acid. If you dry it out it becomes something called gun cotton, which is highly flammable, I grant you, but (turning again to LEFFERTS) doesn't really explode. In the liquid form I am using, known as collodion, it does something rather interesting. Watch.

LEFFERTS watches as HYATT puts a few drops of red ink into the beaker of nitrocellulose solution, which goes bright red. He then dips a wooden ball, suspended on a thread, into the liquid. When he pulls the ball out again, it is coated with a beautiful glossy red plastic, which quickly hardens. The transformation has the required effect on LEFFERTS.

LEFFERTS
Incredible. Can I touch it?

HYATT
(Looking pleased) Yes — well, er, no, it needs to dry a bit more. But here are some I made earlier.

LEFFERTS
(Handling the artificial pool balls, knocking them together in his hands) So you've solved it.
What's the problem? Is it still flammable?

LEFFERTS takes the cigar out of his mouth and gingerly pokes it at the pool ball, which smolders and then bursts into flame. HYATT deftly picks the flaming ball out of LEFFERTS's hands and throws it out of the window.

HYATT
Well, yes, they are flammable. That's not ideal, of course. In fact there have been some reports that when the balls collide at high speed they can spontaneously ignite.
But the real problem is the sound: when the balls knock together they just don't sound right.

LEFFERTS
Pah, who the hell cares what they sound like?

HYATT
Oh, they care. I care, too. But that's not what I wanted to talk to you about. Here. Take a look at this. (He takes an object out of a drawer and hands it to **LEFFERTS.**)

LEFFERTS
(Looks at it for a while) An ivory comb. So what?

HYATT
It's not ivory! (Smiling delightedly) Ha! I fooled you. It's a new material made from this same cellulose nitrate that's coating the wooden ball. But, in my new process, the ball isn't needed. I can make whole objects purely from the cellulose nitrate. You just add naphtha, a solvent derived from crude oil, and hey, presto. The process is called plasticization. (He excitedly starts scrabbling about in his drawer.) Here is a hairbrush, and here is a toothbrush, and here is . . . a necklace . . . (Handing them to **LEFFERTS**)

LEFFERTS is silent for a while as he inspects the fake ivory objects.

LEFFERTS
(Quietly) How big is the market for ivory?

HYATT
Big. Very big.

LEFFERTS
What do you need to start producing this . . . what is it called?

HYATT
It's made from cellulose, so I call it celluloid. What do you think?

LEFFERTS
It's fine with me whatever you want to call it. What do you need to start producing celluloid on an industrial scale?

HYATT
Time and money.

NOTES FOR SCENE 2

All of this is factually accurate (even if the dialogue is somewhat approximate). These days it is hard to believe that anyone could make fundamental chemical discoveries in their shed. But in the late nineteenth century, the beginning of the golden age of chemical engineering, a growing understanding of chemistry coincided with entrepreneurial opportunities for making money out of the invention of new materials. It was also easy and cheap to get hold of chemicals, sales of which were mostly unregulated. Many inventors were operating from their homes—and, in the case of Goodyear, from debtors' prison. Once his rubber proved itself, the demand for the protection, comfort, and flexibility of this type of material grew.

The term *plastic* refers to a huge variety of materials, all of which are organic (which is to say they are made of a group of compounds based on carbon), solid, and moldable. Goodyear's rubber was a form of plastic, but it was the invention of wholly synthetic plastics that revolutionized the term. John Wesley Hyatt and his brother set up a lab in their shed to do just this, inspired in part by an advertisement in the *New York Times* that offered $10,000 to anyone who could invent a new material for billiard balls. Hyatt was also financially backed by a syndicate of investors led by Marshal Lefferts, a retired Civil War general. There were complaints about Hyatt's exploding collodion-covered balls from saloon owners, one of whom reported that "every time the balls collided, every man in the room pulled a gun." These days pool and snooker balls are made from a plastic called phenolic resin, and celluloid is only used for making one type of ball: the table tennis ball.

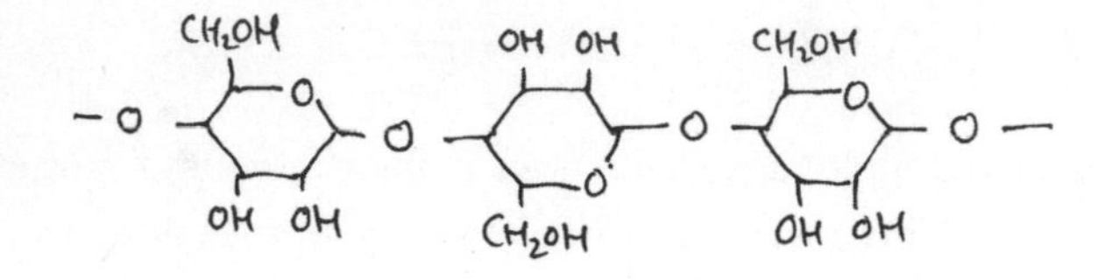

cellulose

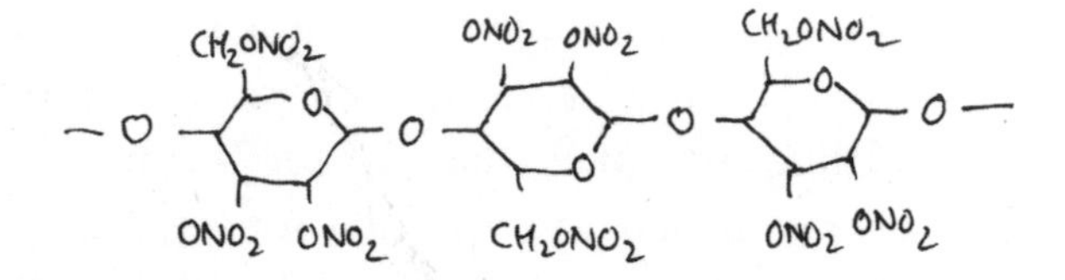

cellulose nitrate

The chemical similarity between celluloid and cellulose, from which paper is made. Both compounds are composed of hexagonal rings made of carbon, hydrogen, and oxygen atoms, conjoined by a single atom.

SCENE 3

Interior: A funeral parlor, San Francisco.

ETHAN'S body is lying naked on an operating table. His clothes, which have just been cut off, are lying on the floor. There are several other new bodies lying on benches around the room, blood dripping from some of them into little pools. There is a strong smell of chemicals, combined with the sweeter, more pungent smell of decay. The **EMBALMER** is cleaning up the blood on **ETHAN**'s body as **BILL** looks on.

BILL

So how long have I got?

EMBALMER

To get his parents here?

BILL nods.

EMBALMER

Three days, under normal circumstances.

BILL

(Through clenched jaws) And what about abnormal circumstances?

EMBALMER

Well, I've got some of the new formaldehyde. With enough of that we can preserve him pretty good, but it's expensive. I can do something cheaper with arsenic, but he won't look the same.

BILL is silent, staring intently at his dead brother, saying nothing.

EMBALMER

So I hear it was the new billiard balls that did him in? The ones made by that New York fella? I was reading about him in the newspaper. A scientist and an inventor, so they say, like Edison, who made those electric lights. But not as successful by all accounts.

BILL

New York? Is he wealthy?

EMBALMER

Must be, I guess . . .

BILL starts walking out.

EMBALMER

Hey, where are you going? What shall I do with your brother's body?

NOTES FOR SCENE 3

In 1869, although the principles of refrigeration were known, it would be another fifty years before refrigerators themselves started to become available. In hot countries there were only two options when someone died: to bury or cremate them, or to embalm them. Embalming methods were based on alcohol or special solutions containing toxic chemicals such as arsenic until 1867, when formaldehyde was discovered by the German chemist August Wilhelm von Hofmann. Unlike previous methods, formaldehyde preserved the tissue in such a way as to give the corpse an almost lifelike appearance, and it soon became the method of choice. Lenin, Kemal Atatürk, and Diana, Princess of Wales, were all embalmed with formaldehyde.

These days, a new technique called plastinization has been developed by Gunther von Hagens. This involves the removal of water and fat (such as lipids) from the body, and their replacement, using a vacuum technique, with silicone rubber and epoxy resin, a hugely versatile material that is used in all sorts of paints and adhesives and flexible products. Like formaldehyde, this produces a lifelike appearance, but because of the stiffness of the plastics used, the bodies can be set into lifelike poses. An exhibition of these preserved and posed bodies, *Body Worlds,* has been touring the world since 1995 and has been seen by millions.

SCENE 4

Interior: Courtroom in New York City, some years later.

HYATT is being questioned about his patent rights on the new celluloid plastic, from which his company is making a lot of money, manufacturing everything from combs and hairbrushes to cutlery handles and even dentures. The **LAWYER** questioning him has been hired by Daniel Spill, an English inventor who claims to have invented a similar plastic, Xylonite, one year earlier. **GENERAL LEFFERTS, HYATT**'s financial backer, is in the front row of a largely empty courtroom, listening to the arguments.

LAWYER

You say that you invented celluloid by attempting to create a replacement material for . . . billiard balls?

HYATT

Yes, that is correct. I was using collodion to coat wooden balls, to give the effect of ivory. But I realized that if I could find a way to make the coating into a solid material, I could do away with the wood, and maybe make a material that sounded more like ivory.

LAWYER

Sounded like ivory? Your story seems rather far-fetched, wouldn't you say?

HYATT

How many times do I need to explain! Any billiard player will tell you that the click of the balls is part of the pleasure of the game.

LAWYER

So you deny receiving information in 1869 from London about a material called Xylonite which uses precisely the same process to turn precisely the same material—(Consults his notes)—cellulose nitrate, into a near-identical solid plastic material, using—(Consults his notes again)—using camphor as a solvent? This is the key step, is it not, that you use to turn collodion into what you call celluloid? Are we to believe this is pure coincidence?

HYATT

No! I mean yes! I do deny it. Absolutely. (Going red with exasperation) I found the method entirely on my own.

LAWYER

Whether you did or did not find it on your own is hardly the point, Mr. Hyatt. As you well know. The point is that there is a prior patent protecting the key process that you are employing in your manufacturing business, and that patent belongs to my client, Mr. Daniel Spill of London, England. To whom you have paid no money.

HYATT

Daniel Spill! Hah! He is not an inventor. He's just an opportunist, a businessman, and a bad one at that! He took all his ideas from Alexander Parkes, a true scientist, the inventor of Parkesine. Spill just copied him. And now he wants to make money out of my honest labors. (Turning to the judge, who is not paying attention) It's a disgrace, your honor.

LAWYER

So are we now expected to believe that you were somehow aware of the work of Alexander Parkes, and yet wholly oblivious to the work of Mr. Daniel Spill?

HYATT

What work of Spill? His material doesn't work! If through some technicality I don't own the patent rights to celluloid, then Spill certainly doesn't either. It was Alexander Parkes who made the first plastic. In 1862. Everyone knows that. Parkes just couldn't get it to work right. But I did—not by copying, like Daniel Spill, but through my own work, by systematic experimentation. (Turning to the judge, who appears to be quite bored and is fiddling with his pocket watch) I just want to run my business without being preyed on by financial parasites!

LEFFERTS has been listening intently throughout, but at **HYATT**'s admission that he knew of Parkesine, **LEFFERTS** looks down for a while, contemplating something, then gets up and leaves.

NOTES FOR SCENE 4

Despite the existence of earlier plastic-like materials, celluloid is widely recognized as being the first commercial moldable plastic. At the International Exhibition of 1862, the British metallurgist, chemist, and inventor Alexander Parkes presented a very curious substance to the world, one made of vegetable matter but that was hard, transparent, and plastic. He called it Parkesine. He too had been obsessed with collodion as a potential plastic, but he had never managed to find a suitable solvent with which to turn the nitrocellulose into a moldable material. It was Hyatt's use of camphor, an odious-smelling gum found in wood, that did the trick. It enabled him to make celluloid into an affordable plastic material.

At the same time, in England Daniel Spill had resurrected the Parkes process, filed for more patents, and launched a similar plastic called Xylonite. Although Xylonite failed commercially, Spill decided to sue Hyatt on the grounds that he had a prior patent on the use of camphor as a solvent in the process. The patent dispute with Daniel Spill almost caused Hyatt's business to close. It was the ruling of the judge—that neither Spill nor Hyatt could claim the patent rights to nitrocellulose plastic—which opened up the whole plastics industry to huge competition and innovation.

SCENE 5

Interior: Boudoir of Mary Louise Young in the town of Boulder, Colorado.

MARY LOUISE is a successful businesswoman who owns the town's only shop. As she talks to **BILL**, she sits in front of the mirror, preparing herself for the evening, combing her hair and trying on jewelry.

MARY LOUISE

Oh, Bill, you only want to marry me to get your hands on my money so that you can be off again on your travels. I know what you're up to.

BILL

There's a man I have business with in New York, but I'll be back as soon as it's taken care of.

MARY LOUISE

(Laughs) So it's true! Well, if I marry, I want to marry for love, Bill. I want to walk down the road arm in arm. I want to take out the horse and cart and go to a picnic at Orchard Creek, and have you feed me grapes . . .
(She giggles at the thought of it.)

BILL

A picnic?

MARY LOUISE

Yes, Bill, a picnic. I want to feel respectable and free, that's what I want marriage to bring me. And I want you to go to the dentist. I won't marry anyone who doesn't have teeth, I reckon that's for certain.

MARY LOUISE is trying on various necklaces. **BILL** gets up angrily and rips the necklaces roughly out of her hand and throws them in the corner.

BILL

Why do you care about this crap?

MARY LOUISE

Bill! Stop it. You always get like this when we talk serious.

BILL

It's just plastic, Mary Louise, plastic. It's not real jewelry, just like you are not a real lady. It's fake jewelry for a fake person!

MARY LOUISE

At least I have aspirations, Bill—and standards! If you want me to take your proposal seriously, you know now what I expect from you . . .

NOTES FOR SCENE 5

The celluloid business boomed in the 1870s and the material was molded into a huge variety of shapes, colors, and textures. Importantly, it could be made to closely resemble much more expensive materials such as ivory, ebony, tortoiseshell, and mother of pearl, and the early forms of plastic were used primarily in this way. Its being relatively cheap to make meant that huge profits were possible from the selling of all manner of plastic combs, necklaces, and pearls to the growing middle class, who were thirsty for the material wealth of the rich but couldn't afford it.

SCENE 6

Interior: A dentist's office.

A plain wooden room with a large chair in the middle and several tables with an array of metal instruments. There is a certificate hanging on the wall, which states that **HAROLD CLAY BOLTON** graduated from the Cincinnati School of Dentistry in 1865. There is a single window in the room, which looks out over some scrubland. It's mid-summer, hot and humid.

DENTIST
Sir, please take off your shirt, and sit down here and get comfortable. (Gesturing to the dentist's chair)

BILL
(Sits down without removing his shirt) How much is this going to cost?

DENTIST
I don't know yet. It all depends on what you need.

BILL
I need teeth. It's pretty simple.

DENTIST
Yes, sir, but I need to look in your mouth first to see what kind of dentures are going to work. I am afraid your shirt may get dirty if you keep it on.

BILL
You ain't doing nothing, you're just looking, right?

DENTIST
Yes, but . . .

BILL
So do it.

DENTIST
I need to take a mold of your gums with this material. (Showing **BILL** some powdered plaster of Paris) And then, depending on how many extra teeth you need, I can use either rubber or this new, rather exciting material, which feels much more comfortable in the mouth.

BILL
I don't care. I just want it to work.

DENTIST
Oh, this new celluloid definitely works. It's marvelously easy to mold and—

BILL
What?!

DENTIST
Celluloid. It's very new, very modern, utterly soft but also . . . hard, if you know what I mean. Which is ideal for our purposes. Everyone is using—(Breaks off as he sees that **BILL** is getting angry) Sir . . . ? Did I say something wrong?

BILL
Goddamn it! Is there no place safe from that damn stuff!

DENTIST
But, sir, plastic really is the best stuff, and so comfortable in the mouth . . . (Following **BILL**, who has gotten up and is walking toward the door) Sir, I don't understand, what's the matter? (Putting his hand on **BILL**'s arm)

BILL angrily pulls away, takes out his gun, and points it at the **DENTIST.**

BILL
I'll tell you what the matter is: you are the matter! (Wagging his gun at the equipment and dental materials) You are all the matter!

NOTES FOR SCENE 6

Oddly, Hyatt did try to build a celluloid business out of creating plastic dentures, but celluloid wasn't suited to that application, mostly because the fake teeth warped in the heat and tasted strongly of the camphor used in their production. Their competition wasn't much better, though, made of rubber and tasting of sulfur. Denture wearers had to wait until the twentieth century for acrylic plastics to give them a much more pleasant, neutral-tasting, and "natural" feel.

SCENE 7

Interior: Hyatt's office, New York.

GEORGE EASTMAN, a camera manufacturer, has come to visit **HYATT** in his office, which is a glass-partitioned corner on the second floor of his celluloid factory.

HYATT

. . . so I believe we could create a camera body that would be much more light-proof than your wooden boxes since they would be made in one piece, and also much less heavy than the metal equivalent.

EASTMAN

I didn't come here to talk to you about the cameras.

HYATT

No?

EASTMAN

No. (**EASTMAN** is quiet for a while. With his back to **HYATT**, he watches the processes going on in the factory below.) How thin can you make celluloid?

HYATT

Thin? Well, I started the business by coating things, if that's what you want.

EASTMAN

(Turning to face **HYATT**, clearly having made up his mind about something) How much do you know about photographic plates?

HYATT

Not much . . . They are made of glass, aren't they?

EASTMAN

Yes, that's right: glass that's been coated in a light-sensitive gel.

HYATT

So . . . you want to use celluloid instead of the gel?

EASTMAN

(Looking mischievous) I want to use celluloid instead of the glass.

HYATT

(Trying to work out why) Hmm . . . so the photographic plates are less easily broken?

EASTMAN

Do you know how many glass plates a photographer can carry, along with all the other equipment he needs?

HYATT shakes his head.

EASTMAN

Ten, maybe fifteen maximum. You practically need a pack animal to carry it all, it's so heavy and cumbersome. Or at least a servant or two — the whole thing is very expensive, a rich man's game.

HYATT

You think plastic photographic plates will make it cheaper?

EASTMAN

I want to turn photography into something everyone can do. So cheap and easy that you could take a camera to a birthday party, or a picnic, or on holiday, or —

HYATT
To the beach!

EASTMAN
Precisely! To do that, we need to make the camera smaller and lighter. But crucially I need to get rid of the heavy glass plates. (Looking seriously at **HYATT**) I have developed such a camera. The trick is to put the photographic emulsion on to a long flexible strip. That way twenty or thirty pictures can fit rolled up in a tiny canister. I am calling it a Kodak camera, and everyone will be able to afford one. I will bring photography to the whole world!

HYATT
So this flexible strip—you have this technology already?

EASTMAN
Well, no. We've been using paper, but it doesn't really work.

HYATT
So, what, you want to use celluloid instead?

EASTMAN
Is it possible?

NOTES FOR SCENE 7

Glass was an excellent material for photographic plates, being both transparent and chemically inert. But the plates were heavy, cumbersome, and expensive, limiting photography to the professionals and to the rich. Celluloid film was developed by George Eastman as a replacement material for glass plates, and was as central to the photographic revolution as his invention of the compact Kodak cameras. By changing from glass plates to a flexible film of celluloid, which could be rolled up, he made the camera much smaller, lighter, cheaper, and simpler. He brought photography to the masses, and in making the camera portable and cheap enough to be used informally, he created a new way of sharing family memories through photography. We now live in a time when buying a roll of film is a rarity for most of us, since the technology has been largely replaced by digital technologies. Nevertheless, the invention of celluloid photographic film was a pivotal moment in visual culture.

SCENE 8

Interior: Hyatt's Office, New York.
A few years later.

It is past midnight. All the lights in the factory are off except one upstairs in **HYATT**'s office, where **HYATT** is fiddling with a strange apparatus. **HYATT** hears a noise and looks up.

HYATT

Who's there? (**HYATT** returns to working, but then hears another noise.) Hello . . . ? Is there anyone there . . . ? Is that you, Betty . . . ?

The door handle of his office twists slowly, and the door swings open. For a moment no one can be seen, but then the figure of **BILL** appears out of the gloom. He is drunk.

BILL

Well, lookee who's here.

HYATT

Who are you? Are you the nightwatchman? Please get out, and do not disturb me again.

BILL

No, I ain't the nightwatchman, but I have been watching. Watching you.

HYATT

What do you mean? Get out. (Getting up) Get out, do you hear?

BILL

No, you ain't ordering me around. As a matter of fact, it's me who's going to order you around.

(Gets out his gun and points it at **HYATT**)
Sit down.

HYATT
I haven't got any money here, if that's what you want. It's all at the bank. It's taken there every day.

BILL
I said, *sit down.*

HYATT
Who are you?

BILL
You killed my brother. So I guess I am going to return the favor and kill you. That sounds fair, doesn't it? So that makes me . . . your executioner.

HYATT
What are you talking about? I haven't killed anyone in my life. There must be some mistake.

BILL
Ain't no mistake. You made the billiard balls that killed him. It's taken me a while to track you down, nigh on ten years it's been since he got shot . . . But here I am.

HYATT
Yes, I did hear something about someone being shot in a saloon while playing pool with my new balls, but it was an accident. It wasn't my fault. I wasn't there!

BILL
It was your fault! It's all your fault! *Shut up!* I'm going to put a stop to all this nonsense. (Gestures to the factory) It ain't natural—and that's why my brother died. You messed with nature . . . putting your stupid plastic material everywhere, making them believe it's valuable like ivory, making money out of women's desire for trinkets, making fools of them. But not me. You ain't making a fool out of me, with your stupid plastic teeth. Someone has to put a stop to this. And it's going to be me.

HYATT

Please! Please don't kill me. Please. Listen to me, please. This material, this plastic you hate, is about to do more for you and your kind than it ever did for anyone else. It is about to immortalize your way of life! It may even turn you into a kind of god – I've seen it!

BILL

What are you talking about? More crap!

HYATT

The problem of making it photographically sensitive has been solved! Moving pictures, haven't you seen them? Stories told on the silver screen. Heroes like yourself, cowboys fighting, winning the West! Everyone is queuing up to see them in the cities. All because of this wonderful flexible transparent material – it couldn't be done with anything else. Storytelling will never be the same again.
Look, I've got a projector here, I have just been trying to feed in the film. Let me show it to you.

BILL

No, it's nonsense, it's all . . .

A light appears behind **BILL**, and the noise of a man's footsteps. The **NIGHTWATCHMAN** appears, carrying a lantern.

NIGHTWATCHMAN

Is everything okay, Mr. Hyatt? I heard shouting.

BILL runs off, knocking the **NIGHTWATCHMAN** to the ground as he makes his escape. The lantern smashes and the naked flame ignites a bucket of discarded celluloid film. **HYATT** and the **NIGHTWATCHMAN** try to put the fire out, but with so much flammable celluloid material on the worktops and in boxes nearby it quite quickly gets out of control. They make their escape, and can only watch from the road as the whole factory is destroyed by the fire.

NOTES FOR SCENE 8

The invention of the roll of film, made possible by the use of celluloid plastic, led directly to the technology of motion pictures. The idea that a picture could be made to "move" by sequentially showing small changes in the image had been known for hundreds of years, but without a flexible transparent material, the only way it could be made to work was using the rotating cylinder of a zoetrope. Celluloid changed everything, allowing a sequence of photographs to be taken on a roll of film and then played back fast enough for the picture to appear to move. This not only allowed a longer sequence of motion to be shown than with the zoetrope, but the moving image could be projected, and so the experience could be shared by the whole audience of a theater. This was the key insight of the Lumière brothers and led to the establishment of the cinema.

The picture on the previous page, of the Wild Bunch, a gang of train robbers led by Butch Cassidy, was taken in Fort Worth, Texas, in 1900. The exploits of the gang typify what we think of as the Wild West, a time of outlaws and violence, which carried on alongside the development of all sorts of modern technologies, such as trains, cars, planes, and of course plastics. The gang's exploits would no doubt have receded into obscurity were it not for the 1969 motion picture starring Paul Newman as Butch Cassidy and Robert Redford as the Sundance Kid. This movie was shot on celluloid film stock, and like many Westerns it immortalized (and romanticized) a way of life that long predated it.

The plastics that followed celluloid, such as Bakelite, nylon, vinyl, and silicone, built on its creative power and have also had an important impact on our cultural psyche. Bakelite became a moldable replacement for wood at a time when the telephone, radio, and television were being invented and needed a new material to embody their modernity. Nylon's sleekness took on the fashion industry, replaced silk as the material for women's stockings, and then spawned a new family of fabrics, such as Lycra and PVC, as well as a group of materials called elastomers, without which all our clothes would be baggy and our pants would fall down. Vinyl changed music, how we recorded it and how we listened to it, and along the way it created rock stars. And silicone—well, silicone turned imagination into reality by creating a plastic form of surgery.

Without plastics, *Butch Cassidy and the Sundance Kid*—and all other movies—would never have existed; neither would the cinema matinee, nor the cinema itself, and our visual culture would be very different indeed. So although I am no fan of excessive plastic packaging, I hope I've shown that, if there is one place a plastic candy wrapper should feel safe and appreciated, it is most definitely in a cinema.

FIN.

7 INVISIBLE

IN 2001, WHILE TRAVELING the country roads of Andalucía in Spain, I experienced a mesmerizing visual effect. I was driving through one of the region's many olive plantations. As the trees that lined the road rushed past, I found myself catching glimpses of the groves moving repeatedly into perfect alignment, flickering like an old silent film. It was as if the ancient olive trees were performing a magic trick for me, in order to relieve the boredom and sticky heat of the journey. These brief snapshots, of line after line of trees stretching seemingly to infinity, were addictive. I watched the road, and then the trick, then the road, then the trick, then I hit a tractor.

To this day I have no idea how it got in front of me. When I slammed on the brakes I was launched out of my seat toward the windscreen. As I smashed my way into it, I remember the moment of contact with the glass, a sudden and intimate freeze-frame as it cracked around me. It felt like hitting a wall of transparent ginger snaps.

Sand is a mixture of tiny bits of stone that have fallen off larger bits of rock as a result of the wind and the waves and other wear and tear that stones have to put up with. If you take a close look at a handful of sand you will find that a lot of these bits of stone are made of quartz, a crystal form of silicon dioxide. There is a lot of quartz in the world because the two most abundant chemical elements in the Earth's crust are oxygen and silicon, which react together to form silicon dioxide molecules (SiO_2). A quartz crystal is just a regular arrangement of these SiO_2 molecules, in the same way that an ice crystal is a regular arrangement of H_2O molecules or iron is a regular arrangement of iron atoms.

Heating up quartz gives the SiO_2 molecules energy and they vibrate, but until they reach a certain temperature they won't have enough energy to break the bonds that hold them to their neighbors. This is the essence of being a solid. If you keep heating them, though, their vibrations will eventually reach a critical value—their melting point—at which they have enough energy to break those bonds and jump around quite chaotically, becoming liquid SiO_2. H_2O molecules do the same thing when ice crystals are melted, becoming liquid water. But there is one very important difference between the two.

The difference is that when the liquid water is cooled, as we all know, crystals re-form with ease and create ice again. It is almost impossible to stop this happening, in fact: from the ice that jams up your freezer compartment to the snow that covers mountains, all are made from liquid water that has refrozen into ice crystals. It is the symmetrical pattern of these H_2O molecules that accounts for the delicate patterns of snowflakes. You can melt and freeze water repeatedly, and the crystals will re-form. With SiO_2 things are different. When this liquid cools down, the SiO_2 molecules find it very difficult to form a crystal again. It's almost as if they can't quite remember how to do it: which molecule goes where, who should be next to whom, appears to be a difficult problem for the SiO_2 molecules. As the liquid gets cooler, the SiO_2 molecules have less and less energy, reducing their ability to move around, which compounds the problem: it gets even harder for them to get to the right position in the crystal structure. The result is a solid material that has the molecular structure of a chaotic liquid: a glass.

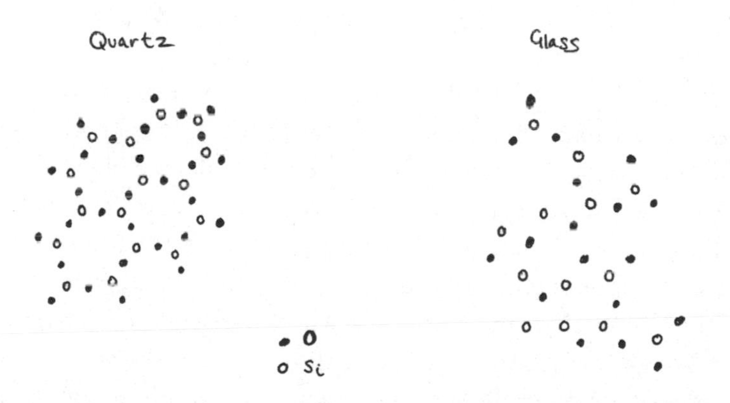

A difference between the crystalline form of silica, a quartz crystal, and the amorphous form of glass.

Since failing to form a crystal is all you need to do to make glass, you'd have thought it would be rather easy. But it's not. Light a fire on the sands of a desert and, with a lot of wind to fan the flames, you may be able to get it hot enough for the sand to start to melt and become a translucent sticky liquid. When this liquid cools, it hardens and indeed becomes glass. But glass made this way will most certainly be full of bits of sand that didn't melt. It will be brown and flaky and will soon fall apart, becoming part of the desert again.

There are two problems with this approach. The first is that most sand doesn't contain the right combination of minerals to make good glass: the brown color is a dreaded sign in chemistry, a clue that you have a mixture of impurities. It is the same with paints: random combinations of colors don't yield pure results; instead you get brownish-gray hues. While some additives, so-called fluxes, such as sodium carbonate, will encourage the formation of glass, most will not. Unfortunately, despite being mainly quartz, sand is also made up of whatever the wind blows in its direction. The second problem is that even if the sand has the right chemical composition, the temperatures needed to melt it are around 1200°C, much hotter than any normal fire, which tends to be in the region of 700–800°C.

A lightning bolt will do the job, though. When one of these strikes the desert it creates temperatures in excess of 10,000°C which are easily high enough to melt the sand, creating shafts of glass called fulgurites. (The word comes from the Latin *fulgur,* which means "thunderbolt"). These glass staffs of charred matter look uncannily like the images of thunderbolts that the gods of thunder, such as the Norse god Thor, were said to hurl in anger. They are surprisingly light, and this is because they are hollow. Although rough on the outside, inside is a smooth, hollow tube, formed when the lightning bolt vaporizes the sand it first encounters. As the heat conducts outward from this entry hole, it melts the sand into a smooth coating for the tube. Further out the temperatures are only high enough to fuse together the sand particles, making their edges rather rough. The colors of fulgurites reflect the composition of the sand in which they are formed, varying from gray-black to translucent if created in a quartz desert. They can be up to fifteen meters long and are fragile, since much of their bulk is made of lightly fused sand. Until recently, they were thought of only as strange curiosities. However, because they trap bubbles of air inside themselves when they form, ancient fulgurites provide scientists studying global warming with a handy record of the desert climates of previous eras.

Fulgurites found in the Libyan Desert.

In one part of the Libyan Desert, there is an area of exceptionally pure white sand, comprised almost entirely of quartz. Search this part of the desert and you may find a rare form of glass that looks nothing like a scruffy fulgurite but which has instead the jewel-like clarity of modern glass. A piece of this desert glass forms the centerpiece of a decorative scarab found on the mummified body of Tutankhamun. We know that this desert glass was not made by the ancient Egyptians because it has recently been established that it is twenty-six million years old. The only glass we know like it is Trinitite glass, the glass formed at the site of the Trinity nuclear bomb test in 1945 at White Sands, Nevada. Given that there was no nuclear bomb in the Libyan Desert twenty-six million years ago, the current theory is that the extremely high temperatures that would have been needed to create such optically pure glass must have been produced by the high-energy impact of a meteor.

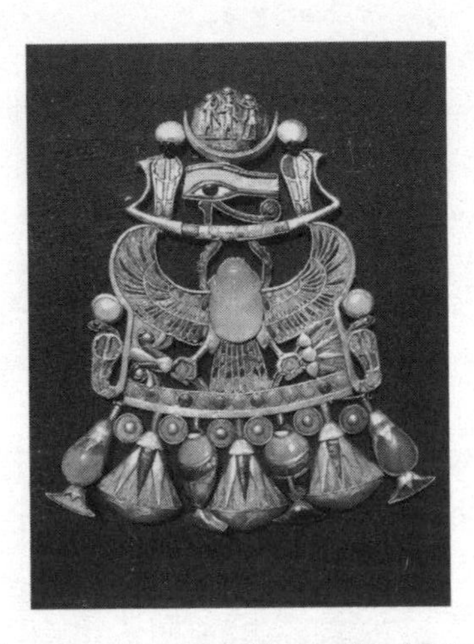

A decorative scarab found on the mummified body of Tutankhamun, with desert glass at its center.

So, without the help of meteor strikes and nuclear explosions, how do you make the kind of glass that we would recognize in our windows, spectacles, and drinking glasses?

Although the Egyptians and the Greeks made advances in glass making, it was the Romans who really brought glass into everyday life. It was they who discovered the beneficial effects of "flux," in their case a mineral fertilizer called natron, which is a naturally occurring form of sodium carbonate. With it, the Romans were able to make transparent glass at a much lower temperature than would be needed to melt pure quartz. In the few locations where the right raw materials and fuel for the high-temperature furnaces were available, they manufactured glass in bulk and then transported it throughout the Roman Empire using their vast trading infrastructure, supplying it to local craftsmen who would turn it into functional objects. None of this was revolutionary; it had been done before, but by making it cheaper, according to Pliny, they put it within the reach of ordinary citizens.

The Roman love of glass as a material is perhaps best demonstrated by their imaginative new uses for it. For instance, they invented the glass window (the word means "wind eye"). Before the Romans, windows were open to the wind, and although these might have wooden shutters or cloth curtains to keep out excessive wind and rain, the idea that a transparent material might be able to provide complete protection was revolutionary. Admittedly their glass windows were small and fused together with lead, because they did not have the technology to make large panes of glass, but they started our obsession with architectural uses of glass, which is still growing today.

Until the development of transparent glass, mirrors were simply metal surfaces, polished to a high shine. The Romans realized that the addition of a layer of transparent glass would protect this metal surface from scratches and corrosion while at the same time allowing them to reduce the metal surface to a thickness of a mere fraction of a millimeter. This dramatically reduced the cost of the mirror and increased its effectiveness and longevity, and remains the basis for almost all mirrors today.

The Romans' innovation in glass technology didn't stop there. Up until the first century AD, almost all glass was crafted into objects by being melted and poured into a mold. This worked well enough for coarse glass objects, but required enormous skill to make anything more delicate. To make a wine glass with thin walls, for instance, required a mold with a thin cavity, but it was hard to get the thick, gloopy molten glass to flow into it. The Romans noticed, though, that solid glass could be made to behave like a plastic if it was hot enough. Using metal pincers, they could pull it into all sorts of shapes before it cooled down too much. They could even blow air into it while it was red hot, and when it cooled they would have a perfect solid bubble. By developing this technique of glass blowing, they were able to blow into existence thin-walled wine glasses, with a delicacy and sophistication that the world had never seen before.

Until this time, drinking vessels had been opaque, made of metal, horn, or ceramic. The appreciation of wine was based solely on the way it tasted. The invention of drinking glasses meant that the color, transparency, and clarity of wine became important, too. We are used to seeing what we drink, but this was new to the Romans, and they loved it.

Although Roman wine glasses were the height of technical and cultural sophistication in their time, compared to modern glasses they were crude. Their main problem was that they were full of bubbles. This was not just an aesthetic problem. It seriously weakened the glass. Whenever a material experiences mechanical stress, which might be caused by anything from being clinked against another glass to being dropped accidentally on the floor, it absorbs the force by dispersing it from atom to atom, reducing the total force that each individual atom has to absorb. Any atom that can't withstand the force being inflicted on it will be ripped from its position in the material, causing a crack. Wherever there is a bubble or crack, the atoms have fewer neighboring atoms to hold them in place and with which to share the force, and so these atoms are more prone to being ripped from position. When a glass smashes, it is because the force is so great that a chain reaction occurs within the material, with the failure of each atom causing the failure of its neighbor. The bigger the force, the smaller the bubble or crack needed to initiate this chain reaction. Or to put it another way, large bubbles in your wine glass mean it won't be able to withstand much impact.

The extreme fragility of glass might explain why glass making took so long to catch on after the Romans, despite their having made so much progress. The Chinese knew how to make glass, and even traded for Roman glass, but they didn't develop it themselves. This is quite surprising given that the Chinese mastery over materials technologies outshone that of the West for a thousand years after the Roman Empire collapsed. The Chinese were experts in paper, wood, ceramic, and metals, but they pretty much ignored glass.

By contrast, in the West, the fashion for wine glasses nurtured a respect and appreciation for glass that ultimately had a profound cultural impact. In Europe, and especially the colder Northern Europe, transparent waterproof glass windowpanes, which let the light in but kept the elements out, were too desirable a technology to ignore. At first, only tiny panes of glass could be made that were of sufficient purity and consistency not to shatter, but these could be knitted together to make larger windows using lead. They could even be colored with glazes. Colored or stained glass windows became a means of expressing wealth and sophistication, changing entirely the architecture of the European cathedral. Over time the artisans making stained glass for cathedrals became as high status as the masons who cut the stone, and in Europe the new art of glazing blossomed.

The disdain for glass in the East lasted all the way up until the nineteenth century. Before then, the Japanese and Chinese relied on paper for the windows of their buildings, a material that worked perfectly well but resulted in a different kind of architecture. The lack of glass technology in the East meant that, despite their technical sophistication, they never invented the telescope nor the microscope, and had access to neither until Western missionaries introduced them. Whether it was the lack of these two crucial optical instruments that prevented the Chinese from capitalizing on their technological superiority and instigating a scientific revolution, as happened in the West in the seventeenth century, is impossible to say. What is certain, though, is that without a telescope you can't see that Jupiter has moons, or that Pluto exists, or make the astronomical measurements that underpin our modern understanding of the universe. Similarly, without the microscope, it is impossible to see cells such as bacteria and to study systematically the microscopic world, which was essential to the development of medicine and engineering.

So why is it that glass has this apparently miraculous property of transparency? How is it that light can travel through this solid material at all, while most other materials will not allow it? After all, glass contains all of the same atoms that make up a handful of sand. Why in the form of sand should they be opaque and in the form of glass transparent and able to bend light?

Glass is made of silicon and oxygen atoms, as well as a few other sorts. Within every atom there is a central nucleus, which contains protons and neutrons, surrounded by varying numbers of electrons. The size of the nucleus and the individual electrons is tiny compared to the overall size of the atom. If an atom were the size of an athletics stadium, the nucleus would be the size of a pea at its center, and the electrons would be the size of grains of sand in the surrounding stands. So within all atoms—and indeed all matter—there is a majority of empty space. This suggests that there should be plenty of room for light to travel through an atom without bumping into either an electron or the nucleus. Which indeed there is. So the real question is not "Why is glass transparent?," but "Why aren't all materials transparent?"

Inside an atomic stadium, to continue the analogy, the electrons are only allowed to inhabit certain parts of the stands. It is as if most of the seats have been removed and there are only certain

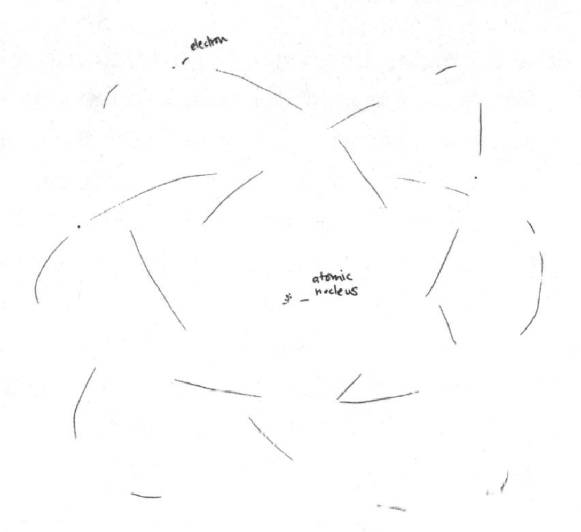

A sketch of an atom showing that it is mostly free space.

rows of seats left, with each electron restricted to its allotted row. If an electron wants to upgrade to a better row, it has to pay more—the currency being energy. When light passes through an atom it provides a burst of energy, and if the amount of energy provided is enough, an electron will use that energy to move into a better seat. In doing so, it absorbs the light, preventing it from passing through the material.

But there is a catch. The energy of the light has to match exactly that required for the electron to move from its seat to a seat in the available row. If it's too small, or to put it another way, if there are no seats available in the row above (i.e., the energy required to get to them is too large), then the electron cannot upgrade and the light will not be absorbed. This idea of electrons not being able to move between rows (or energy states, as they are called) unless the energy exactly matches is the theory that governs the atomic world, called quantum mechanics. The gaps between rows correspond to specific quantities of energy, or quanta. The way these quanta are arranged in glass is such that moving to a free row requires much more energy than is available in visible light. Consequently, visible light does not have enough energy to allow the electrons to upgrade their seats and has no choice but to pass straight through the atoms. This is why glass is transparent. Higher-energy light, on the other hand, such as UV light, can upgrade the electrons in glass to the better seats, and so glass is opaque to UV light. This is why you can't get a suntan through glass, since the UV light never reaches you. Opaque materials like wood and stone effectively have lots of cheap seats available and so visible light and UV are easily absorbed by them.

Even if light is not absorbed by glass, moving through the interior of an atom still affects it, slowing it down until it emerges from the other side of the glass, when it speeds up again. If the light strikes the glass at an angle, different parts of the light will enter it and emerge from it at different instants, forcing them to travel momentarily at slightly different speeds. This momentary difference is what causes the light to bend, or refract, and this is what makes an optical lens possible, with the curvature of the glass resulting in different angles of refraction at different points along its surface. Controlling the curvature of the glass means we can magnify images, which allows us to make telescopes and microscopes, and, for those of us who wear glasses, to see.

It also, and perhaps more fundamentally, turns light into a subject for experimentation itself. Through the centuries all glass makers must have noticed that glass could create mini-rainbows on the wall as sunlight hit it at particular angles, but no one could explain the cause, except to state the blindingly obvious, which is that the colors were somehow being generated within the glass. It wasn't until 1666 that Isaac Newton realized that what was blindingly obvious was blindingly wrong and came up with the real explanation.

Newton's moment of genius was to notice that a glass prism not only turned "white" light into a mixture of colors, but could also reverse the process. From this, he deduced that all of the colors created by a piece of glass were already in the light in the first place. They had traveled all the way from the sun as a ray of mixed light, only to be split up into their constituent colors when they hit the glass. The same thing would happen if they hit a drop of water, too, since this was also transparent. At a stroke, Newton had for the first time in history managed to explain the main features of the rainbow.

The satisfying explanation of an atmospheric phenomenon using a laboratory experiment showed the power of scientific reasoning. It also showcased the role of glass as laboratory accomplice in unraveling the mysteries of the world. This role was not limited to optics. Chemistry was transformed by glass perhaps more than any other discipline. You only have to go to any chemistry lab to see that the transparency and inertness of the material make it perfect for mixing chemicals and monitoring what they do. Before the glass test tube was born, chemical reactions were performed in opaque beakers, so it was hard to see what was happening. With glass, and especially with a new glass called Pyrex that was immune to thermal shock, chemistry as a systematic discipline really got going.

Pyrex is a glass with boron oxide added to the mix. This is another molecule that, like silicon dioxide, finds it hard to form crystals. More importantly, as an additive it counteracts the tendency of glass to expand when heated or contract when cooled. When different parts of a piece of glass are at different temperatures, expanding and contracting at different rates, stresses build up within the material as the different parts of the glass strain against one another. These stresses cause cracks to grow and ultimately shatter the glass. If this happens in a vessel containing boiling sulfuric acid, such a failure can maim or kill. The discovery of borosilicate glass (Pyrex is a trade name) put a stop to thermal expansion and the stresses associated with it. It released chemists to heat and cool their experiments as they wished, to concentrate on chemistry and not the potential dangers of thermal shock.

Chemists were also able to bend glass tubes within the laboratory with the aid of only a blowtorch and to construct complex chemical equipment such as distillation vessels and gas-tight containers much more easily. Gases could be collected, liquids controlled, chemical reactions allowed to do whatever they liked. Glass equipment is the workhorse of the chemist's world—so much so that every professional chemical lab has a glass blower in residence. How many Nobel Prizes did this material make possible? How many modern inventions started life in a test tube?

Whether the relationship between glass technology and the seventeenth-century scientific revolution really is a simple case of cause and effect is an open question. It seems more likely that glass was a necessary condition rather than the reason for it. However, there is no doubt that glass was largely ignored in the East for a thousand years. And during this time, glass revolutionized one of Europe's most treasured customs.

While glass had been used by the rich to drink wine for hundreds of years, most beers until the nineteenth century were drunk from opaque vessels such as ceramic, pewter, or wooden mugs. Since most people couldn't see the color of the liquid they were drinking, it presumably didn't matter much what these beers looked like, only what they tasted like. Mostly, they were dark brown and murky brews. Then in 1840 in Bohemia, a region in what is now the Czech Republic, a method to mass-produce glass was developed, and it became cheap enough to serve beer to everyone in glasses. As a result people could see for the first time what their beer looked like, and they often did not like what they saw: the so-called top-fermented brews were variable not just in their taste, but in their color and clarity too. Not ten years later, a new beer was developed in Pilsen using bottom-fermenting yeast. It was lighter in color, it was clear and golden, it had bubbles like champagne—it was lager. This was a beer to be drunk with the eyes as much as with the mouth, and these light golden lagers have continued in this tradition ever since, being designed to be served in a glass. How ironic, then, that so much lager is drunk from an opaque metal can, meaning that the only beer uniquely identifiable for its visual appearance is the epitome of opaqueness, a beer in the old pre-glass tradition, Guinness.

The move to serving beer in glasses had another unexpected side effect. According to the UK government, more than five thousand people are attacked with glasses and bottles every year, costing the health service more than £2 billion to surgically repair the injured. Although many alternative plastic materials for serving beer in bars and pubs have been tried, materials which are both transparent and tough, they have never gained acceptance. Drinking beer from a plastic cup is a completely different experience to drinking from a glass. Not only does plastic taste different, but it also has a lower thermal conductivity, a property that makes it feel warmer than glass, reducing the satisfaction of drinking an ice-cold beer. Plastic is also much softer than glass, so plastic beer cups soon become tarnished, scratched, and opaque. This masks the clarity of the beer but it also affects our perception of the cleanliness of the vessel. One of the great attractions of glass is that its shiny bright appearance makes it seem clean even if it isn't, a collective deception we all accept in order to avoid thinking too much about using the same glass that was in a stranger's mouth perhaps only an hour before. Creating plastics that are hard enough to withstand scratching is a major goal for materials science. Such a discovery could then be used to make lighter windows for airplanes, trains, and cars, and lighter screens for mobile phones, but so far it seems completely out of reach. In the meantime, we have found another solution to the problem: instead of replacing glass, just make it safer.

Such glass is called toughened glass and was invented by the automobile industry in order to reduce the injuries caused by shards of glass in car crashes. The scientific origin of toughened glass is to be found in a famous curiosity of the 1640s known as Prince Rupert's drops. These are teardrop-shaped pieces of glass that can withstand intense pressure at their rounded end, but if they incur the slightest damage at the tail end they will explode. Prince Rupert's drops are very simple to make: all you need to do is drop a small piece of molten glass into water. The extreme and rapid cooling of the outside of the droplet puts the surface layers of the glass into a state of mechanical compression. All of the glass here is pushing in against itself, and as a result cracks find it very hard to form since the compression stress is always pushing the sides of the crack back together. This has the effect of toughening the outside of the glass to the point where the glass drop can, incredibly, withstand even a hammer blow.

However, to maintain this compression stress in its surface, the laws of physics require an equal and opposite "tensile" stress in its interior. As a result, the atoms in the middle of the drop are in a state of high tension: they are all being pulled away from one another. They are, in effect, like a small explosion waiting to go off. If the surface compression becomes ever so slightly unbalanced, which can be achieved by making a small indent in the tail of the drop, a chain reaction courses through the whole material as all the atoms in tension snap back into place—and the material explodes into countless tiny shards. These shards are still sharp enough to cut, but they are small enough not to do any major damage. Getting windscreens to behave in a similar way was just a matter of finding a method for cooling the outside of the glass fast enough to create the state of compression found in Prince Rupert's drops. The material that resulted has saved countless lives in car crashes, where it dissolves characteristically into millions of tiny shards.

Over the years, glass has been made even safer. The windscreen I hit in Spain was made of the latest generation of safety glass, called laminated glass. I knew this because, although it shattered in the manner of Prince Rupert's drops, the fragments of the windscreen were held together in a single piece as we both made our journey across the bonnet of the car and onto the tarmac.

This new generation of toughened glass has a layer of plastic in its middle, which acts as a glue keeping all the shards of glass together. This layer, known as a laminate, is also the secret behind bulletproof glass, which is essentially the same technology but with several layers of plastic embedded at intervals within the glass. When a bullet hits this material, the outermost layer of glass shatters, absorbing some of the bullet's energy and blunting its tip. The bullet must then push the glass shards through the layer of plastic beneath it, which flows like tough treacle, thus spreading the force over a wider area than the point of impact. No sooner has it got through this layer than the blunted bullet encounters another layer of glass, and the process starts all over again.

The more layers of glass and plastic there are, the more energy the bulletproof glass can absorb. One layer of laminate will stop a 9-millimeter pistol bullet, three layers will stop a .44 Magnum pistol bullet, eight layers will stop a person with an AK-47 rifle from killing you. Of course, there is little point having a glass bulletproof window if you can't see through it, so the real challenge lies not so much in the layering of the materials but in matching the refractive index of the plastic with that of the glass, so that light is not bent too much as it travels from one to the other.

This technologically sophisticated laminated safety glass is more expensive to produce, but it is increasingly a price we are prepared to pay to enjoy its benefits. The material is popping up all over the place, not just in cars but across modern cities as they become more and more like glass palaces. In the summer of 2011 there were riots in many city centers of the UK. Viewing the TV footage I couldn't help but notice a difference between these and other such riots I had seen in the past: occasionally the rioters were unable to break the windows of shops by throwing a brick at them because many businesses had installed toughened safety glass. This trend is likely to increase, the shops using glass not just to present their wares but to protect them too. This same laminated glass has been proposed as the material for new safety beer glasses, which would put an end to the use of glass as a weapon in bars and pubs.

It is now impossible to imagine a modern city without glass. On the one hand, we expect our buildings to protect us from the weather: this is what they are for, after all. And yet, faced with a prospective new home or place of work, one of the first questions people ask is: how much natural light is there? The glass buildings that rise every day in a modern city are the engineering answer to these conflicting desires: to be at once sheltered from the wind, the cold, and the rain, to be secure from intrusion and thieves, but not to live in darkness. The life we lead indoors, which for many of us is the vast majority of our time, is made light and delightful by glass. Glass windows have come to signify that we are open for business, and that the business will be honest and open—a shop without a shop window is practically not a shop at all.

This material is also instrumental in how we view ourselves. You may be able to see yourself reflected in a shiny metal surface or a pond, but for most of us it is the glass mirror that has become the final, intimate arbiter of our self-image. Even photographs and video representations are mediated through the glass lens.

It is often said that there are very few places left on Earth that have yet to be discovered. But those who say this are usually referring to the places that exist at the human scale. Take a magnifying glass to any part of your house and you will find a whole new world to explore. Use a powerful microscope and you will find another, complete with a zoo of living organisms of the most fantastic nature. Alternatively, use a telescope and a whole universe of possibilities will open up before you. Ants build cities at their scale, and bacteria build cities at their scale. There is nothing special about our scale, about our cities, about our civilization, except that we have a material that allows us to transcend our scale—that material is glass.

Yet we have no great love for the material that has made this possible. People do not tend to wax lyrical about glass in the way that they do about, say, a wooden floor or a cast-iron railway station. We do not run our hands down the latest double-glazed panel and admire the sensuality of this material. Maybe this is because in its purest form it is a featureless material: smooth, transparent, and cold. These are not human qualities. People tend to relate more to colored, intricate, delicate, or simply misshapen glass, but this is rarely functional. The most effective glass, the stuff we build our modern cities from, is flat, thick, and perfectly transparent, but it is the least likable, the least knowable: the most invisible.

For all its considerable importance in our history and our lives, glass has somehow failed to win our affections. When we break a glass window it is shocking, annoying, and painful in the case of my Spanish car accident; but we do not feel that we have broken something that is intrinsically valuable. In these situations we are worried for ourselves, but as for the glass itself, it can be replaced. Perhaps it is because we look through it rather than at it that glass has not become part of the treasured fabric of our lives. The very thing that we value it for has also disqualified it from our affections: it is inert and invisible, not just optically, but culturally.

THE FIRST TIME I went to art class, the teacher, a man called Barrington, told us that everything we could see was made of atoms. Everything. And that if we could understand that, we could begin to be artists. The room went quiet. He asked for questions, but all of us were struck dumb, wondering if we were in the right class. He continued his introduction to art by holding up his pencil and proceeding to draw a perfect circle on the piece of paper he had taped to the wall. There was general excitement and sighs of relief from the assembled pupils. Perhaps we were in an art class after all.

"I've transferred atoms from the pencil to the paper," he observed. He then gave us a speech about the wonders of graphite as a material for artistic expression. "It is important to note," he said, "that although diamond is culturally revered as the superior form of carbon, it is in fact incapable of deep expression, and unlike graphite no good art can come from diamond." What he subsequently thought of Damien Hirst's diamond-encrusted skull, *For the Love of God,* valued at £50 million, I can only guess.

But in describing the relationship between the two forms of carbon, diamond and graphite, as a rivalry, he was certainly correct. The battle between dark, expressive, utilitarian graphite and sublime, cool, hard, glinting diamond has been raging since antiquity. In terms of cultural value, diamond has long been the winner, but that may be set to change. A new understanding of graphite's inner structure has made it a source of wonder.

Thirty years after my introduction to graphite by my art teacher, I met Professor Andre Geim, one of the world's foremost carbon experts, in his fluorescent-lit office on the third floor of the Physics Department of Manchester University. I wish I could say that, like Barrington, he too used only graphite to express himself, but when he opened his desk drawer I saw that it was full of ballpoint pens and whiteboard markers. In his thick Russian accent, Andre said, "There is no such thing as a perfect circle, Mark," leaving me a little unsure as to whether he had understood the point of my story. Then he fished out of the drawer a small red leather presentation case and said, "Have a look at that while I make the coffee."

Inside the case was a disk of pure gold the size of a biscuit, decorated with the relief portrait of a man. As I weighed the heavy disk in my hand I found it almost obscenely metallic: gold is the full-fat cream of the metal world. I was taken aback by the decadence of the material. The man depicted was Alfred Nobel, and the inscription on the medal announced to the world that Andre Geim's team had received the 2010 Nobel Prize for physics, for his groundbreaking work on graphene, a two-dimensional version of graphite and a marvel of the materials world. As I waited for Andre to return with the coffee, I pondered his cryptic answer. Perhaps he was suggesting that although his last ten years of research on carbon had been circular, he had not ended up where he started.

Carbon is a light atom with six protons and usually six neutrons in its nucleus. Sometimes it contains eight neutrons, but in this form, known as carbon-14, the atomic nucleus is unstable, and so the element falls apart through radioactive decay. Because the rate of this decay is consistent over long periods of time, and because this form of carbon finds its way into many materials, measuring its presence in a material allows us to work out that material's age. This scientific method, known as carbon dating, has thrown more light than any other on our ancient past. The true ages of Stonehenge, the Turin Shroud, and the Dead Sea Scrolls have all been revealed by this form of carbon.

Radioactivity aside, the nucleus plays a back-seat role in carbon. In terms of all of its other properties and behavior, it is the six electrons that surround and shield the nucleus that are important. Two of these electrons are deeply embedded in an inner core near the nucleus and play no role in the atom's chemical life—its interaction with other elements. This leaves four electrons, which form its outermost layer, that are active. It is these four electrons that make the difference between the graphite of a pencil and the diamond of an engagement ring.

The simplest thing a carbon atom can do is share each of these four electrons with another carbon atom, forming four chemical bonds. This solves the problem of its active four electrons: each electron is partnered off with a corresponding electron, belonging to another carbon atom. The crystal structure produced is extremely rigid. It is a diamond.

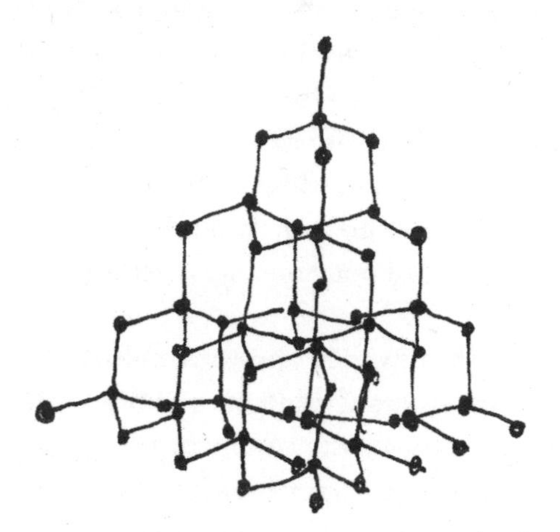

The crystal structure of diamond.

The biggest diamond yet discovered is located in the Milky Way in the constellation of Serpens Cauda, where it is orbiting a pulsar star called PSR J1719–1438. It is an entire planet five times the size of Earth. Diamonds on Earth are minuscule by comparison. The biggest yet found is the size of a football. Extracted from the Cullinan mine in South Africa, it was eventually presented to King Edward VII in 1907 on his birthday and is now part of the crown jewels of the British monarchy. This diamond was formed far below the surface of the Earth at a depth of approximately three hundred kilometers, where, over the course of billions of years, the high temperatures and pressures converted a largish-sized carbon rock into the huge diamond. The diamond was then most likely carried to the surface of our planet during a volcanic eruption, where it lay inert and undisturbed for millions of years until it was discovered a mile underground.

I was constantly being dragged to museums as a child, to the national this or that, and without exception I was bored in all of them. I tried to do what the adults did and walk around in ponderous silence or ruminate in front of a painting or a sculpture, but it didn't work for me. I got nothing out of it as far as I could fathom. It was very different when we visited the crown jewels. I was entranced from the moment I set foot inside. This was a true Aladdin's cave. The gold and jewels seemed to speak a fundamental language to me, more fundamental than art, more primitive. A feeling akin to religious devotion came over me. Looking back, I think that this experience was not a reveling in wealth but was my reaction to being exposed for the first time to a pure form of materiality. There was a huge scrum of people in front of the Great Star of Africa diamond (as the largest gem from the Cullinan diamond was called after it was cut to shape). A mere glimpse of this diamond was enough for me never to forget it, even from my viewpoint under the armpit of a giant man wearing a damp lumberjack shirt and behind a tutting Indian woman. The presence of the Indian lady was appropriate, I later found out from my dad's encyclopedia, since India was the sole source of diamonds until the mid-eighteenth century, when they were discovered in other parts of the world, most notably South Africa.

Each diamond is, in fact, a single crystal. In a typical diamond there are about a million billion billion atoms (1,000,000,000,000,000,000,000,000), perfectly arranged and assembled into this pyramidal structure. And it is this structure that accounts for its remarkable properties. In this formation, the electrons are locked into an extremely stable state, and this is what gives it its legendary strength. It is also transparent, but with an unusually high optical dispersion, which means that it splits light that enters it into its constituent colors, giving it its bright rainbow sparkle.

The combination of extreme hardness and optical luster makes diamonds almost flawless as gemstones. Because of their hardness, virtually nothing can scratch them, and so they keep their perfectly faceted shape and pristine sparkle not just throughout the lifetime of the wearer but throughout the lifetime of a civilization—through rain or shine, whether worn in a sandstorm, hacking through a jungle, or just doing the washing up. Even in antiquity diamond was known to be the hardest material in the world. The word *diamond* is derived from the Greek *adamas,* meaning "unalterable" or "unbreakable."

Transporting the Cullinan diamond back to Britain posed an enormous security challenge for its owners, since the discovery of the largest ever rough diamond had been widely reported in the newspapers. Notorious criminals like Adam Worth, the inspiration for Sherlock Holmes's nemesis, Moriarty, who had already managed to steal a whole shipment of diamonds, were perceived to be a real threat. In the end, a plan worthy of Sherlock Holmes was hatched and executed. A decoy stone was dispatched on a steamboat under high security while the real one was sent by post in a plain brown paper box. The ruse worked precisely because of another remarkable attribute of diamond: being composed solely of carbon, it is extremely light. The entire Cullinan diamond would have weighed little more than half a kilogram.

Adam Worth was not alone. As the wealthy were acquiring large diamonds at an ever increasing rate, a new type of occupation was being born at this time: that of cat burglar. Diamond's lightness and high value meant that stealing a diamond, even one the size of a marble, could allow you to retire for life, and once stolen it was essentially untraceable. (Contrast this to stealing Andre Geim's gold medal, which would have gained me a few thousand pounds at most when melted down.) This new type of jewel thief was imbued with the virtues of diamond itself: elegant, sophisticated, and unadulterated. In films such as *To Catch a Thief* and *The Pink Panther,* diamonds play the role of a princess cruelly imprisoned. Upstanding members of society by day, cat burglars by night, their rescuers were played by film stars like Cary Grant and David Niven. In these films, a diamond robbery is portrayed as a noble act. The diamond thief is light on his or her feet and requires only a black catsuit and a knowledge of sophisticated stately homes and combination safes located behind paintings. In contrast, the stealing of cash or gold from a bank or train was depicted as a grubby crime, often carried out by brutish and greedy men.

Unlike gold, diamonds have never been part of the world's monetary system, despite their financial value. They are not a liquid asset—and quite literally so: they cannot be melted down and, in this way, commodified. Large diamond gemstones have no use except to arouse wonder and awe and, most importantly, to affirm status. Before the twentieth century only the truly rich could afford them. But the growing wealth of the European middle classes provided a tempting new market for diamond miners. The problem faced by the company DeBeers, which in 1902 controlled 90 percent of the world's diamond production, was how to sell to this much bigger market without devaluing the gems in the process. They managed it through a cunning marketing campaign: by concocting the phrase "Diamonds are forever," they invented the idea of the diamond engagement ring as the only true way to express everlasting love. Anyone who wished to convince their lover of the truth of their feelings needed to buy one, and the more expensive the diamond, the truer the feelings expressed. The marketing campaign took off spectacularly, catapulting a diamond into millions of households and culminating in a James Bond movie, accompanied by a Shirley Bassey / John Barry song, that enshrined the new social role of the diamond as the embodiment of romantic love.

But diamonds are not forever, at least on the surface of this planet. It is, in fact, diamond's sibling structure, graphite, that is the more stable form, and so all diamonds, including the Great Star of Africa in the Tower of London, are actually turning slowly into graphite. This is distressing news for anyone who owns a diamond, although they can be reassured that it will take billions of years before they see an appreciable degradation of their gems.

The structure of graphite is radically different from diamond. It consists of planes of carbon atoms connected in a hexagonal pattern. Each plane is an extremely strong and stable structure, and the bonds between the carbon atoms are stronger than those in diamond—which is surprising, given that graphite is so weak that it is used as a lubricant and as lead in pencils.

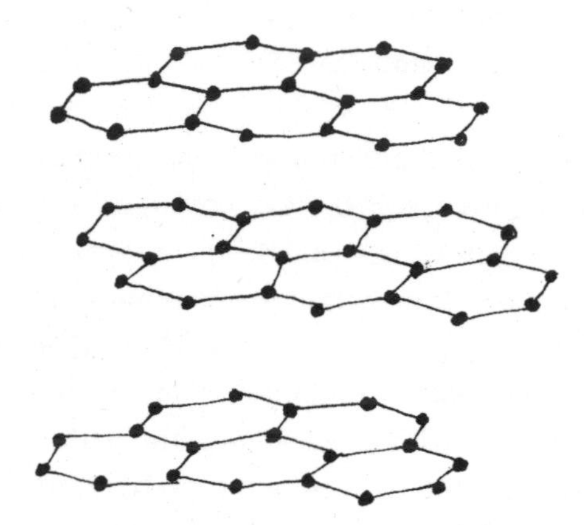

The crystal structure of graphite.

The conundrum can be explained by noting that *within* the graphite layers each carbon atom has three neighbors with which it shares its four electrons. In the diamond structure, each carbon atom shares its four electrons with four atoms. This gives the individual graphite layers a different electronic structure and stronger chemical bonding than diamond. The flip side, though, is that each atom in graphite has no electrons left over to form strong bonds *between* its layers. Instead, these layers are held together by the universal glue of the material world, a weak set of forces generated by fluctuations in the electric field of molecules, called van der Waals forces. This is the same force that makes Blu-Tack sticky. The upshot is that when graphite is put under stress, it is the weak van der Waals forces that break first, making graphite very soft. This is how a pencil works: as you press it on the paper you break the van der Waals bonds and layers of graphite slide across one another, depositing themselves on the page. If it weren't for the weak van der Waals bonds, graphite would be stronger than diamond. This was one of the starting points for Andre Geim's team.

Take a look at the graphite of a pencil and you will see that it is dark gray and shiny like a metal. For thousands of years it was mistaken for lead and was referred to as "plumbago," or "black lead," hence the use of the term "lead" to refer to the graphite used in a pencil. The confusion is understandable since they are both soft metals (although these days we call graphite a semi-metal). Plumbago mines became more and more valuable as new uses were found for graphite, such as the discovery that it was the perfect material to cast cannon and musket balls. In seventeenth- and eighteenth-century Britain the material became so expensive that thieves took to digging secret tunnels into the mines, or working in the mines and secreting the plumbago about their person. As the price increased so did the smuggling and criminal activity until an act of Parliament was passed in 1752 that made the stealing of graphite from a mine a felony punishable by a year's hard labor or seven years' transportation to Australia. By 1800 graphite was such big business that armed guards were posted at the entrance to plumbago mines.

The reason why graphite is metallic while diamond is not is also its hexagonal atomic structure. As we have seen, in the diamond structure, all four electrons in each carbon atom are partnered up with a corresponding electron. In this way, all atoms in the lattice are strongly held in a bond, and there are no "free" electrons. This is the reason why diamonds do not conduct electricity, because there are no electrons free to move within the structure to carry the electric current. In the graphite structure, on the other hand, the outer electrons do not just bond with a counterpart electron in a neighboring atom but rather form a sea of electrons within the material. This has several effects, one of which is to allow graphite to conduct electricity, since the electrons can move around like fluid. Graphite was used by Edison for the first light bulb filaments because it also has a high melting point, which allows it to glow white hot without melting when a high current passes through it. Meanwhile, the sea of electrons also acts as an electromagnetic trampoline for light, and this reflection of light is what makes it appear shiny like other metals. This neat explanation of graphite's metallic properties is not what won Andre Geim's team a Nobel Prize, though. It was merely their starting point.

All forms of life on Earth are based on carbon, and although these types of carbon appear very different from graphite, they can easily be converted into its hexagonal structure through burning: wood turns into charcoal when heated; bread becomes burnt toast; we too become black and charred when exposed to a fire. None of these processes produce pure shiny graphite, since the hexagonal layers of carbon are not densely packed but are jumbled up. But there is a vast spectrum of black sooty materials, which all have one thing in common: they contain carbon in its most stable form—hexagonal sheets.

In the nineteenth century, another form of black sooty carbon rose to ascendancy: coal. The hexagonal planes of carbon atoms in coal are formed not through heat, as with a burnt piece of toast, but through geological processes acting on dead plant matter over millions of years. Coal starts off as a form of peat, but as heat and pressure act upon it, depending on the exact conditions, it is transformed into lignite, then bituminous coal, then anthracite coal, and finally graphite. What happens as it undergoes this transformation is that, step by step, it loses the volatile compounds, containing nitrogen, sulfur, and oxygen, that are present in the original plant matter, becoming in the process a purer and purer form of carbon. As the pure hexagonal layers get formed, so the material takes on a more metallic shine, which can be seen particularly clearly in the black mirror facets of some coals such as anthracite. Nevertheless, coal is very rarely a pure form of carbon, which is why it can be quite smelly when it burns.

The type of coal most revered for its aesthetic appeal is that derived from fossilized monkey puzzle trees. It is hard, can be carved and polished to a brilliant finish, and has a beautiful dark black luster. It is sometimes called black amber because it has similar triboelectric properties to amber: the ability to generate static charge and make hair stand on end. We know it more commonly as jet. It was made fashionable in Britain in the nineteenth century by Queen Victoria, who mourned the death of her consort Prince Albert by wearing black clothes and jet jewelry for the rest of her life. There was subsequently such a popular demand for jet from the rest of the British Empire that overnight the population of the Yorkshire town of Whitby, where Bram Stoker later wrote his Gothic masterpiece *Dracula,* stopped using the large local deposits of jet for fuel and became famous for producing the jewelry of lament and sorrow.

The idea that diamond has anything in common with coal or graphite was pure fantasy until early chemists started to investigate what happened when you heat it. Antoine Lavoisier did just this in 1772 and found that diamond burns when it gets red hot, leaving nothing. Nothing at all. It just seems to disappear into thin air. This experiment was extremely surprising. Other gemstones, such as ruby and sapphire, were found to be impervious to red heat or even white heat; they did not burn. But diamond, the king of gems, seemed to have an Achilles' heel. What Lavoisier did next makes my heart sing, such is the elegance of the experiment. He heated diamond in a vacuum so that there, with no air to react with the diamond, it might survive to higher temperatures. It's one of those experiments that is easy to propose but much harder to carry out, especially in the eighteenth century, when vacuums themselves were not so easy to produce. What happened next astounded Lavoisier: the diamond still wasn't impervious to red heat, but this time it turned into pure graphite—proof that these two materials were indeed made of the same stuff, carbon.

Armed with this knowledge, Lavoisier and countless others across Europe searched for a way to reverse the process, to turn graphite into diamond. Vast wealth would be the reward for anyone who could do it, and the race was on. But the task was a formidable one. All materials prefer to change from less stable to more stable structures, and because the diamond structure is less stable than graphite's, it requires very high temperatures and pressures to persuade it to change in the opposite direction. These conditions exist inside the Earth's crust, but it still takes billions of years to grow a big diamond crystal. Simulating the conditions in a laboratory is extremely difficult. Claim after claim was made and then retracted. None of the scientists involved got massively rich overnight, which some said was further evidence of their failure. Others suspected that those who did achieve the feat of transformation kept quiet and got rich slowly.

Whatever the truth, it wasn't until 1953 that there was reliable documented evidence of such a transformation being achieved. Now the synthetic diamond industry is indeed big business, but it does not compete head to head with the natural diamond jewelry industry. There are a few reasons for this. The first is that although the industrial process has been mastered to the extent that small synthetic diamonds can be produced more cheaply than mining real ones, they are mostly colored and flawed, since the accelerated process of making them introduces defects which color the diamonds. In fact, the majority of these diamonds are used in the mining industry, where they embroider drills and cutting tools, not for aesthetic effect but to enable them to cut through granite and other hard rocks. Secondly, much of the value of diamonds is derived from their authenticity. It is important when proposing marriage that the diamond you offer, although physically identical to a synthetic one, was forged in the depths of the Earth a billion years ago. Thirdly, if you are the ultra-rational sort of person who does not care about the natural history of a gemstone, then buying a synthetic diamond is still a very expensive way to embellish your loved one. There are much cheaper lustrous substitutes that will glitter and dazzle and still fool anyone except a diamond expert, such as cubic zirconia crystals or even glass.

However, the preeminence of natural diamond, in its fight for supremacy with graphite, was to take another blow when it was found that it was no longer the hardest known material. In 1967 it was discovered that there is a third way of arranging carbon atoms that produces an even harder substance than diamond. The structure is based on graphite's hexagonal planes but modified to be three-dimensional. This structure, called lonsdaleite, is thought to be 58 percent harder than diamond, although it exists in such small quantities that it is hard to test. The first sample was found in the Canyon Diablo meteorite, where the intense heat and pressure of impact transformed graphite into lonsdaleite. An engagement ring has never been made from lonsdaleite, since the types of meteorite impacts that create it are extremely rare and produce only tiny crystals, but the discovery of this third structure of carbon led, perhaps inevitably, to the question of whether yet further carbon structures could exist, in addition to the cubic structure of diamond, the hexagons of coal, jet, charcoal, and graphite, and the three-dimensional hexagonal structure of lonsdaleite. Soon, another synthetic one joined the list, thanks to the aircraft industry.

Early aircraft were made of wood because it is light and stiff. Indeed, one of the fastest aircraft in the Second World War was a wooden airplane called the Mosquito. Making airframes out of wood is problematic because it's hard to join into a defect-free structure. So as the scale of aircraft engineers' ambitions increased they turned instead to a light metal called aluminum. But even aluminum is not super-light, and nagging away at the back of many engineers' minds was the hope that there might somehow be a material that was stronger and lighter even than aluminum. It didn't seem to exist, so in 1963 engineers from the Royal Aircraft Establishment in Farnborough decided to invent one.

Carbon fiber, as they named it, was made by spinning graphite into a fiber. By rolling sheets of this material up, with the fibers running lengthwise, they could take advantage of the huge strength and stiffness within the sheets. The weakness, as with pure graphite, still lay in the material's structural dependence on van der Waals forces, but this was overcome by encasing the fibers in an epoxy glue. A new material was born: carbon fiber composite.

Although this material would, in the end, displace aluminum in the building of aircraft (the recent Boeing Dreamliner is 70 percent carbon fiber composite), it took a long time for carbon fiber to prove itself worthy of the aircraft industry. Sports equipment manufacturers, however, took a liking to it immediately. It transformed the performance of racket sports so quickly that those who stuck to traditional materials such as wood and aluminum were quickly outclassed. I remember vividly that moment in my life when my friend James appeared on the tennis court one day wielding a composite tennis racket with the characteristic black-on-black checkered patterning of carbon fiber. Before we played the match, he allowed me to experience its extreme lightness and power with a few practice shots—and then swapped back and thrashed me. There is something highly unnerving about playing an opponent who has a racket that is half the weight and has twice the power of yours. "You carbon be serious!" I exclaimed. It didn't help.

Soon the material was changing any and every sport that required low-weight and high-power materials—in other words, pretty much all of them. Bicycle racing was transformed in the 1990s when engineers started to produce bikes with ever more aerodynamic shapes using carbon fiber structures. The development of these bikes probably reached its zenith in Chris Boardman's classic sporting rivalry with Graeme Obree to beat "The Hour" record: the competition that seeks to determine the furthest a human being can travel in one hour under their own power. In the 1990s both cyclists were able to smash the world record and then each other's records repeatedly with the help of ever more sophisticated carbon fiber bicycles. In 1996 Chris Boardman rode 56.375 kilometers in one hour and provoked an outcry from the International Cycling Union. They promptly banned the use of these new carbon fiber–inspired designs, so worried were they by how radically the bikes would change the nature of the sport.

Formula One took the opposite approach to the innovation offered by carbon fiber, with constant changes in the rules forcing further innovation in materials design. Indeed, mastery of technology is integral to the sport, and success is achieved as much through engineering advances as it is through the skill of the driver. Meanwhile, carbon fiber plays a role even in the sport of running. More and more disabled athletes are using carbon fiber transtibial artificial limbs. In 2008 the International Association of Athletics Federations tried to prevent such athletes from competing against able-bodied athletes on the grounds that the carbon fiber legs gave them an unfair advantage. However, this ruling was overturned by the Court for Arbitration of Sport, and in 2011 the athlete Oscar Pistorius competed as part of the able-bodied South African World Championship 4×400-meter relay team, which won a silver medal. Carbon fiber is set to become a big part of athletics unless the athletics federations take the same approach as the cycling federations.

The huge success of carbon fiber composite inspired engineers to imagine its use on the grandest possible scale: was this material strong enough to achieve a longstanding dream, that of building an elevator into space? The Space Elevator, also known by its aliases Sky Hook, Heavenly Ladder, and Cosmic Funicular, would be a structure linking a point on the equator to a satellite in geostationary orbit directly above it. If a space elevator could be constructed, it would democratize space travel at a stroke, allowing people and cargo to be transported into space with ease and with an almost negligible energy cost. The concept, which was developed in 1960 by a Russian engineer, Yuri Artsutanov, would require the construction of a thirty-six-thousand-kilometer-long cable connecting a satellite to a ship floating in the ocean at the Earth's equator. All studies indicate that the idea is mechanically feasible but requires the cable to be made from a material with an extraordinarily high strength-to-weight ratio. The reason why weight comes into it, as with any cable structure, is that it must first be able to hold its own weight without snapping. At thirty-six thousand kilometers long, you would need a material so strong that a single thread of it could be used to lift an elephant. In practice even the best carbon fiber thread could only lift a cat. But this is because it is full of defects. Theoretical calculations make clear that if a completely pure carbon fiber could be engineered, then its strength would be much higher, exceeding the strength of diamond. The search was on to find a way to make such a material.

A clue to how this might be done came with the discovery of a fourth carbon structure, one that was found in the most unlikely of places: the flame of a candle. In 1985 Professor Harry Kroto and his team discovered that inside a candle flame carbon atoms were miraculously self-assembling in groups of exactly sixty atoms to form super-molecules of carbon. The molecules looked like giant footballs and were nicknamed "buckyballs" after the architect Buckminster Fuller, who had designed geodesic domes with the same hexagonal structure. Kroto's team received the 1996 Nobel Prize for chemistry for this discovery, and also woke everyone up to the fact that the microscopic world might contain a whole zoo of other carbon structures that had never been seen before.

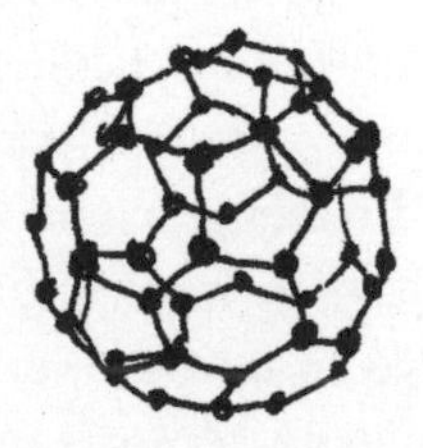

The molecular structure of "buckyballs."

Almost overnight carbon became one of the sexiest topics in materials science, and soon another type of carbon emerged, a carbon that could form tubes that are only a few nanometers wide. Despite the complexity of their molecular architecture, these carbon nanotubes had a peculiar property: they could self-assemble. They needed no outside help in order to form these complex shapes, nor did they need high-tech equipment. They could do it in the smoke of a candle. It was a moment on a par with the discovery of microscopic bacteria; the world suddenly seemed a much more complex and extraordinary place than we had imagined. It wasn't just living organisms that could self-assemble into complex structures; the non-living world could do it too. An obsession with the production and examination of nanoscale molecules gripped the world, and nanotechnology became fashionable.

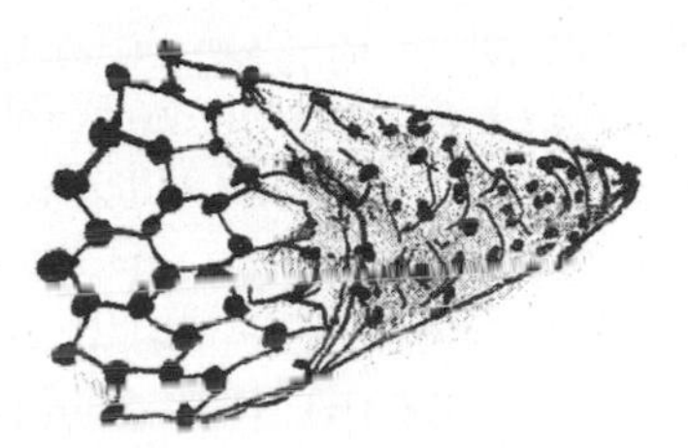

The molecular structure of carbon nanotubes.

Carbon nanotubes are like miniature carbon fibers except that they have no weak van der Waals bonding. They were found to have the highest strength-to-weight ratio of any material on the planet, which meant that they might be strong enough to build a space elevator. Problem solved? Well, not quite. Carbon nanotubes are, at most, a few hundred nanometers in length, but they would need to be meters in length to be of use. Currently there are hundreds of nanotechnology research teams around the world working to solve this problem. Andre Geim's team, though, is not one of these.

Andre's team asked a simpler question: if all of these new forms of carbon were based on the hexagonal structure of graphite, and graphite was full of these layers of hexagonal carbon, then why wasn't graphite a wonder material too? Answer: because the sheets slip over one another too easily, so the material is very weak. But then what if there were only one sheet of hexagonal carbon? What would that material be like?

When Andre Geim came back with the coffee, I was still holding his gold Nobel Prize medal in my hand. I felt vaguely guilty despite the fact that it was he who had given it to me to look at. He put the coffee down, took the medal out of my hand, and replaced it with a lump of pure graphite from the Plumbago Mines of Cumbria. He had, he said, obtained the graphite direct from the mine, which was just up the road, geographically speaking, from his office at Manchester University. Then he showed me how his research group had made a single sheet of hexagonal carbon.

He took a piece of sticky tape and stuck it on to the lump of graphite. When he removed it a thin wafer of brightly metallic graphite was stuck to the tape. He then took another piece of the tape and stuck that to the thin wafer, and then peeled it back. Now the wafer had been split into two parts. Doing this four or five times created yet thinner wafers of graphite. Finally he announced that he had made some graphite that was one atom thick. I looked at the piece of sticky tape he was holding. It had a few black smudges on it, and, not wanting to downplay the significance of it, I peered intently. "Of course," he said with a smile, "you cannot see it. At that scale it is transparent." I nodded exaggeratedly as he took me next door to the microscope, which would allow us to see these atomic layers of graphite.

Andre's team didn't get the Nobel Prize for making a single layer of graphite. They got the Nobel Prize for demonstrating that these single layers of graphite had properties that were extraordinary even by nanotechnology standards—so extraordinary that they merited their own name as a new material: graphene.

Just for starters, graphene is the thinnest, strongest, and stiffest material in the world; it conducts heat faster than any other known material; it can carry more electricity, faster and with less resistance, than any other material; it allows Klein tunneling, an exotic quantum effect in which electrons within the material can tunnel through barriers as if they were not there. All this means that the material has the potential to be an electronic powerhouse, possibly replacing silicon chips at the heart of all computation and communication. Its extreme thinness, transparency, strength, and electronic properties mean also that it may end up being the material of choice for touch interfaces of the future, not just the touch screens we are used to but perhaps bringing touch sensitivity to whole objects and even buildings. But its most intriguing claim to fame is that it is a two-dimensional material. This doesn't mean it has no thickness, but rather that it cannot be made any thicker or thinner and be the same material. This is what Andre's team showed: add an extra layer of carbon to graphene and it goes

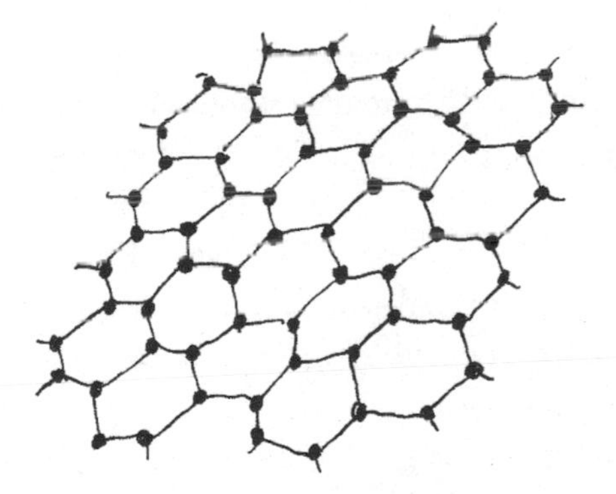

The molecular structure of graphene.

back to being graphite, take a layer away and the material does not exist at all.

Although my art teacher, Barrington, didn't know it when he claimed that graphite was a superior form of carbon to diamond, he was right in almost every technical sense. He was right also about the importance of the atomic nature of graphite. Graphene is the atomically thin building block of graphite. It is what you sometimes deposit on your paper as you use a pencil. It can be used solely as an expressive artistic material. But it is much more than that: this material and its rolled-up version in the form of nanotubes are going to be an important part of our future world, from the smallest scale to the very largest, from electronics, to cars, to airplanes, rockets, and even – who knows? – to space elevators.

So has graphite, by giving birth to graphene, finally outshone diamond? Is it the unexpected winner of this age-old rivalry? It's a bit too early to tell, but it seems doubtful to me. Although it appears likely that graphene will usher in a new age of engineering, and indeed scientists and engineers are in love with this material already, this may not give it high status in the world at large. Diamonds may not be the hardest, strongest material anymore, and we know that they will not last forever, but they still represent those qualities to most people. They are still the rock that romantically binds lovers everywhere. The association of diamond with true love may originally have been thanks to a PR campaign, but it is no less real to us now.

Graphene, on the other hand, may be functionally better than diamond, but it doesn't sparkle and is virtually invisible, extremely thin, and two-dimensional – not the kinds of qualities that anyone wants to associate their love with. My guess is that until the PR companies discover graphene, the cubic crystal structure of carbon will continue to be a girl's best friend.

9 REFINED

IN JANUARY 1962 the Miodownik family was gearing up to celebrate the marriage of my father, Peter Miodownik, to his fiancée, Kathleen. Wedding plans were in place, friends had been invited, religious instruction for the ceremony between this Jewish man and Catholic woman was ongoing, nerves were jangling, free love may or may not have been practiced, but wedding presents for the young couple were definitely being ordered, one of which was a bone china tea set.

It was delivered to my parents' house in a wooden crate from the Harrods department store. Once the cups and saucers had been removed from their bed of sawdust, they were washed and placed on the kitchen draining board. Here they got the first view of their home: a bare but large kitchen in the suburbs of London. One of the tea cups slid from the sink top on to the floor, but it bounced off the linoleum rather than smashing, much to the joy of the happy couple, who grinned at each other in amazement. A good omen, they decided, and so it turned out: the cups served them their whole marriage. Fifty years later, the one I have in the picture of me on my roof is the sole survivor.

In the early days, the porcelain tea cups had to share a kitchen cupboard with some wooden cups that my mum had brought with her from Ireland. This must have filled them with horror. Wood has a rustic appeal, of course; it is a beautiful and natural material, and its organic simplicity is appealing to those who long for a more rural life. But as a material to drink out of it is hard to defend. It has a strong taste and absorbs other flavors into its pores with ease, distorting the taste of the next drink.

Metal cups were also knocking around the kitchen at that time. Apparently they were from a camping set, and had been brought into the kitchen because the newlyweds had few others. But metal is not much better for drinking tea out of than wood. We put metal cutlery in our mouths all the time, preferring it to other materials because its stiffness and strength allow forks and spoons to be thin and sleek without bending or snapping. Crucially, their shininess and smoothness make it easy to detect whether they have been completely cleaned since they were last in another's mouth. But the material conducts heat too well to be used for hot drinks.

It also sounds brash and loud, an acoustic signature that doesn't match the sophisticated flavors of tea.

Some plastic cups joined the household when my brothers and I were born. Like most objects designed for children, they are colorful and robust, and this suits the drinks they contain, which tend to be much sweeter and fruitier than tea. The feeling of soft plastic in the mouth, meanwhile, is warm, comforting, and safe. They look jolly and sweet, the material mirroring the state of infancy. It would be appropriate if plastic juice cups grew up to be ceramic tea cups as they got older, becoming stronger, stiffer, and more distinguished. But sadly what happens to plastic cups is that they die young, structurally degraded by the UV rays of the sun. Every picnic takes years off a plastic cup's life. Eventually they go yellow and brittle, and finally fall apart.

Ceramics, on the other hand, are impervious to UV degradation or chemical attack. They resist scratching better than any other material, too. Oils, fats, and most stains just bounce off them. Tannins and a few other molecules do stick to them, but acid or bleach cleans them fairly easily. As a result, ceramics keep their looks for a very long time. In fact, if it wasn't for the crack in my cup, which runs from its lip to its handle and has become stained with tannins, it would look pretty much the same as it did fifty years ago. There are very few things you can say that for. Paper cups may seem sustainable because paper is recyclable, but the wax coating required to make them waterproof makes this almost impossible. For real sustainability, we must look to ceramics.

Practicalities aside, there is a social stigma attached to serving tea in paper, plastic, metal, or pretty much any material other than ceramic. Tea drinking is about so much more than ingesting fluid: it is a social ritual and a celebration of certain ideals. Ceramic cups are an essential part of this ritual — an essential part, therefore, of a civilized home.

The story of how ceramics got their high status dates from a long time ago, before paper, before plastics, before glass, and before metal. It all began when humans started putting the clay from river beds into fires in the realization that they could transform it. It didn't just dry out. No, something else took place that turned the soft squidgy clay into a rigid new material that had almost all the qualities of stone. It was hard and strong and could be shaped into storage and collecting vessels for grain and water. Without these vessels, agriculture and settlements would have been impossible, and civilization as we know it would never have got off the ground. Roughly ten thousand years later, these vessels came to be known as pots and this simple species of ceramic as pottery.

But these early ceramics were not really like stone. They were fragile, easily broken, dusty, and porous (because at a microscale their skin was full of holes). Terra cotta and earthenware are modern relatives of these early ceramics. They are beautifully simple to make but are still terribly weak. There have been countless occasions when I have put one of these terra cotta dishes, usually bought on holiday, into the oven containing some casserole only for it to emerge an hour later cracked and leaking. Of all places, the oven is where ceramics should be comfortable, because that's where they were formed, but terra cotta fails time and time again. The reason is that liquid seeps into its pores and then expands into steam when heated, turning the pore into an exploding micro-crack that eventually links up with other micro-cracks like tributaries of a river, and finally erupts on the surface of the terra cotta dish, spelling an end not just to the dish but, as often as not, to the meal within it, too.

Unlike metals, plastics, or glass, ceramics cannot be melted and poured. Or rather, we don't have the materials that can withstand the temperatures required to contain such liquids. Ceramics are made from the same stuff as mountains, rocks, and stone, whose liquid form is the lava and magma of the Earth. But even if lava could be captured and poured into a mold, it would not form a strong ceramic—certainly not one that you would recognize or make a cup from. What forms is, of course, volcanic rock, which is full of holes and imperfections. It takes millions of years of heat and pressure deep inside the Earth to turn such stuff into the so-called igneous rocks and stones that make up mountains. For these reasons, attempts to make artificial substitutes for rock either use chemical reactions, which is how cement and concrete work, or, in the case of pottery, involve heating up clay in a furnace, not to melt it, but instead to take advantage of a very unusual property of crystals.

Clay is a mixture of finely powdered minerals and water. Like sand, these mineral powders are the result of the eroding action of the wind and water on rocks, and are in fact tiny crystals. Clays are formed often in river beds, where these eroded minerals are washed down from mountains, settle on the river bed, and form a squidgy, soft dough. Different mixtures of minerals result in different kinds of clay. In the case of terra cotta, the crystals are usually a mixture of quartz, alumina, and rust, which gives the terra cotta its red color.

When this is heated up, the first thing that happens is that the water evaporates, leaving the tiny crystals aggregated in a kind of sand castle with lots of holes where the water used to be. But at high temperatures something special happens: atoms from one crystal will jump on to another nearby crystal and then back again. The atoms in some crystals, however, do not return to their original position, and gradually bridges of atoms are built between the crystals. Eventually, billions of such bridges are built, and the collection of crystals has become something more like a single continuous mass.

The reason the atoms do this is the same reason why any two chemicals react: within each crystal, all of the atom's electrons are part of a stable chemical bond with its neighbors—they are, as it were, "occupied"—but at the edges and surfaces of the crystal, there are "unoccupied" electrons, ones that have no other atoms to bind to, the equivalent of loose ends. For this reason, all of the atoms in a crystal seek a position within the body of the crystal rather than at its surface; or, put another way, those atoms at the surface of the crystal are unstable, available, and liable to relocate if an appropriate opportunity to do so comes their way.

How firing of ceramics transforms an assembly of small crystals into a physically coherent single material.

Usually, when the crystals are cold, these atoms don't have enough energy to move around and do something about their predicament. But when the temperature is high enough, the atoms can move around: they set about reorganizing themselves so that as few of them as possible are forced to inhabit a position at the surface of the crystal—so that, in fact, there is less surface overall. In doing so, they reshape the crystals to fit together as fully and economically as possible, eliminating the holes between them. Slowly but surely the collection of tiny crystals become a single material. It's not magic, but it is magical.

That's the theory, of course, but the chemistry of some clays makes this easier to do than that of others. The advantage of terra cotta clays is that they are easy to find, and this reshaping process will happen at relatively low temperatures—the temperature of a fire or simple wood furnace will do. This means that making terra cotta requires only a small degree of technical know-how. As a result, whole towns and cities are built from the stuff: the common house brick is essentially a form of terra cotta. The big problem with terra cotta ceramics, though, is that they never get rid of all the holes, and so never become fully dense. This is fine for house bricks, which only need to be fairly strong, and once cemented in place will not be bashed around or heated and cooled repeatedly. But it is a disaster for a cup or a bowl, which will have a thin body but be expected to withstand the rigors of the kitchen. They just don't last: one small knock and the cracks start to grow from the pores and never stop.

It was the potters of the East who solved the problem of fragility and porosity. Their first step was to realize that if earthenware was covered with a particular kind of ash, this ash would transform during firing into a glass coating that would stick to the outside of the pot. This glass skin would seal all the pores on the outside of the earthenware. And by varying the composition and distribution of the glaze, the pots could be colored and decorated. This not only stopped water getting in but it suddenly opened up a whole new aesthetic realm for ceramics.

These days you quite often see this glazed earthenware. There is certainly a lot of it in my kitchen—in the form of the tiles that cover the walls around the sink and cooking surfaces, making them easy to clean and pretty to look at—and it is all over bathrooms and toilets. The use of patterned tiles to cover floors, walls, and even whole buildings is associated particularly with Middle Eastern and Arabic architecture.

While glazing prevents water from getting into fired clay, it doesn't solve the problem of porosity within the body of the ceramic, which is how the cracks start in the first place. So tiles are still relatively weak, as are glazed terra cotta cups and bowls. This problem was also solved by the Chinese, but it involved the creation of a completely new type of ceramic altogether.

Two thousand years ago, while looking for a way to improve their ceramics, the potters of the Eastern Han Dynasty started experimenting not just with different kinds of clays but with clays of their own concoction, mixing into them minerals that might never end up in a river. One such additive was the white mineral kaolin. Why? No one knows. Perhaps it was purely in the spirit of inquiry, perhaps because they liked the color.

No doubt they tried all sorts of different mixtures, but eventually they hit upon a particular combination of kaolin and a few other ingredients, such as the minerals quartz and feldspar, which created a white clay and, when fired, a nice-looking white ceramic. This was no stronger than earthenware, but, unlike any other clay they knew, if they increased the temperature of the furnace to a very hot 1300°C, it did something strange. The clay turned into an almost watery-looking solid: a white ceramic that had a near perfectly smooth surface. It was quite simply the most beautiful ceramic that anyone had ever seen. It was also stronger and tougher than any ceramic had any right to be. It was so strong that cups and bowls could be made that were extremely thin, almost as thin as paper, without compromising their ability to withstand cracks. These cups were so fine that they were translucent. It was porcelain.

This combination of properties—strength, lightness, delicacy, and extraordinary smoothness—made a powerful statement, and the material soon became associated with royalty, projecting an image of their wealth and sophisticated aesthetic taste. But it had another meaning too: because it required deep knowledge and skill to create precisely the right mixture of minerals to make it and to build the furnaces that could generate the high temperatures to fire it, porcelain came to represent the perfect marriage of technical skill and artistic expression. What began as a source of pride for the Han Dynasty soon became a matter of identity, embodying their prowess. From that moment on in Chinese history, different royal dynasties were associated with different types of imperial porcelain.

The dynasties showed off their ceramics by creating incredibly beautiful vases and ceremonial bowls with which they decorated their palaces. But they realized that for their honored guests to really marvel at the translucency and lightness of this new material they needed to feel it as well as see it. Tea drinking provided a perfect way to do so. Serving tea to one's guests in porcelain cups became an expression not just of technical sophistication but of cultural refinement as it grew eventually into a ceremonial ritual.

Chinese porcelain was so superior to any other ceramic that when traders from the Middle East and the West came into contact with it, they immediately realized how valuable it would be as a commodity. They exported not just the porcelain but the tea-drinking ritual as well, which together became the ambassadors of Chinese culture, causing a sensation wherever they went. At this time Europeans were still drinking from wood, pewter, silver, or earthenware. Porcelain was physical evidence of just how much more technically advanced the Chinese were than anyone else in the world. To have a set of porcelain tea cups and to serve the best China tea immediately set you apart. Consequently an enormous trade in this sublime white porcelain, called "white gold" or "china," started up.

The trade became so great that many in Europe realized that if they could learn how to make porcelain themselves, they would become very rich. But no one got close, and the method for making porcelain remained a jealously guarded secret known only to the Chinese, despite the Europeans sending spies to China to find it out. It wasn't for five hundred years, when a man named Johann Friedrich Böttger was imprisoned by the king of Saxony and told that his life depended on discovering it, that a European porcelain was first created.

Böttger was an alchemist, but in 1704, while in prison, he was made to work under the instruction of a man called von Tschirnhaus to carry out a systematic set of experiments using various white minerals in order to create porcelain. The discovery of a local deposit of kaolin proved the turning point. Once they had achieved the high temperatures required, they discovered what the Chinese had known for more than a thousand years.

Böttger chose to prove that he had created porcelain not by serving tea with the new cups but by removing one from the white-hot furnace where it had just been fired at 1350°C and plunging it straight into a bucket of water. Most ceramics would shatter under these extreme circumstances because of the thermal shock; earthenware and pottery would explode. But the toughness and strength of porcelain was so great that it survived intact.[1] The king duly rewarded Böttger and von Tschirnhaus handsomely, because the invention of European porcelain was about to make him exceedingly rich.

After this, scientists and potters from all over Europe began experimenting in order to discover the secret of porcelain for themselves. Industrial espionage was rife, but it still took another fifty years for the British to come up with their own version of porcelain, using local ingredients, called "bone china porcelain." And it was from this material that the tea set given to my parents as a wedding gift was made.

In 1962, then, the year of the announcement of the Miodownik wedding, miners in Cornwall would have set out, as they had each morning for two hundred years, to make their way through the wild ferns of the Cornish hills, scattered with pits and water mills, to the Treviscoe Pit to dig out an unusual deposit of the rare white clay, kaolin. Meanwhile, up the road in a granite mine, other miners would have been extracting stone, including mica, feldspar, and

---

1 Although this story is widely disbelieved, we re-created this experiment in July 2011 for the BBC4 series *Ceramics: How They Work* and confirmed that porcelain does indeed survive the thermal shock of being plunged white hot into water.

quartz. Farmers in the county of Staffordshire and the surrounding counties of Cheshire, Derbyshire, Leicestershire, Warwickshire, Worcestershire, and Shropshire would have been raising livestock, whose bones would eventually be burnt and crushed into a powder. All of these ingredients would have found their way to Stoke-on-Trent, the place where, on a winter's day, my cup and the others in its set were born.

At this time of year, the city would have been thick with smog, created by the hundreds of red-brick bottle-shaped kilns that made this one of the homes of British pottery. The smog in those days would have had a very distinctive sulfurous and slightly acidic smell. Perhaps, as they often did when I lived there in 1987, the clouds would be hanging so low that the smoking chimneys seemed to blend into them, making the town feel unreal, as if it was part of a dream. Inside the factories the air, heated by the kilns, would have been warm, dry, and cozy. Room after room would have been full of benches and mechanical equipment, attended by lines of busy men and women absorbed in their work, making ceramics of all kinds, but mainly plates, saucers, and, of course, tea cups. The activity would have been tremendous, an air of industrious concentration pervading all. And everything would have been being made from one substance. It dominated the factory and left its mark everywhere. The whole place would have been stained with this fine white powder, a mixture of minerals and bone.

The powder itself would have looked utterly unremarkable, and even when water was added and it became a workable clay with the consistency of sticky pastry, it would not seem to have had much more going for it. The cups would have been molded by hand in the Wedgwood factory by a woman who had been doing it all her life. They would have gone from pastry blob to cup very quickly, with the aid of a potter's wheel and the deft hands of such a master craftsperson. They would have then been set on a tray, fragile and wet, with almost no strength, like premature babies.

Without help now, they would have dried, sagged, cracked, and then fallen apart, just as a cup made from mud will do. But instead they would have been whisked off to another part of the factory.

There, a man with enormous blunt fingers and immense dexterity would have quickly built a box called a saggar out of a type of fireclay (clay that can withstand particularly high temperatures and is therefore used to house other clays when they are being fired) and placed them all inside. They would have been carefully arranged and mechanically supported so that none of them touched any other. Then, when all was ready, they would have been sealed into the saggar with a final piece of clay. It would have been dark inside the saggar, and cold and damp with them all still wet and weak.

The next day the saggar, along with about five hundred others, would have been carefully stacked inside one of the bottle-shaped kilns until it was completely full. The kiln would have then been sealed and the coal fires lit underneath it. Protected by the saggar, which would have been exposed to the smoke and the fumes, the cups would have remained pristine white, drying out slowly as the temperature increased over the course of a day until all the water that was binding them together had evaporated. Now would have been the most delicate stage of their birth. At this point, the cups would have been helplessly weak, a cluster of mineral crystals clinging to each other but with nothing between them to act as a glue, while the saggar protected them from the strong currents of super-heated air and smoke which would have otherwise blown them apart.

As the temperature increased further, the minerals' components, the crystals, would have started to morph and change shape. Atoms would have danced from one crystal to another, building bridges between them and rearranging the whole internal architecture of the cup into a single solid mass.

Then, as the temperature increased further still to 1300°C and the whole kiln became white hot, the magic would have started to happen: some of the atoms flowing between their crystals would have turned into a river of glass. Now they were mostly solid, but also part liquid. It would have been as if the cups had blood running through their veins in the form of liquid glass. This liquid would have flowed into all the small pores between the crystals and coated all the surfaces. Now, unlike almost all other types of ceramic, the cups would have felt what it was like to be free of defects.

It would have taken two days for the kiln to cool down enough for it to be opened, but the cups would have been still too hot to be safely removed. Nevertheless, a whole troop of sturdy, burly men, black with soot and wearing three layers of woolen jumpers and coats, would have come in to take them out. A few saggars would have cracked open during the firing, exposing their cups to the smoke and flames of the kiln, a sad ending for them. But the Miodownik cups would have been safe, still entombed in their saggar womb, until this was carefully cracked open, and they were delivered into the world, fully fledged bone china of the most extraordinary kind. They would have been inspected for defects and then, for the final test, rather like slapping a baby's bottom, each one would have been given an expert flick.

The ringing sound of a cup is the clearest and surest way to know whether it is fully formed inside. If there are any defects within it, any holes that were never filled by the river of glass that flowed while they were white hot, then these will absorb some of the sound and prevent it from reverberating. Such a cup sounds dull. But a fully dense cup rings and rings and rings. It is this ring that would have pronounced the Miodownik cups' official acceptance into the world. Tap a terra cotta cup and you will hear almost nothing, or a dull thud at best. But because the porcelain in my cup is fully dense, without any defects, it has kept its fine and delicate shape, despite being paper thin and translucent, for some fifty years, and you can still hear that strength and vitality in its ring.

These cups were used for all the special occasions of the Miodownik family. They were part of the tea service used when my mother's mother visited from Ireland to see her daughter's new home. They were there when the whole family gathered to celebrate the birth of the first Miodownik son, Sean. They were there when the neighbors were invited over to celebrate the Silver Jubilee in 1977 and Uncle Alan secretly drank vodka out of one of them and fell over in the flower beds. They were there on Christmas Day when Opa Miodownik sneezed, covering the fully laden dining table with snot and leading to such a kerfuffle that one of the cups was swept from the table and smashed on the floor. They were there when each of the Miodownik sons got married, except for Sean, who jumped out of an airplane in Hawaii and got married on a beach.

As prized porcelain wedding gifts, these cups only saw the ceremonial side of the Miodownik family. They were only brought out to impress on special occasions. They never participated in daily life: they did not deliver tea in bed, nor to the garden wall by the vegetable patch, nor to the boys playing football. These domestic situations are the province of the mug, a lower-quality glazed stoneware or earthenware cup. These are thick-walled because they are made from too weak a material to survive otherwise. Cheap and cheerful, it is their informal shape and size that makes them so homely. The tea drunk from them is pretty cheap and cheerful too. It is the British national drink, despite being Chinese in origin. Its role is quite different, though, from the tea that was served by the Han Dynasty in order to display its wealth and sophistication. Tea in Britain is made predominantly from a tea bag containing blended, finely ground tea of the cheapest sort. We like it to look dark, brown, and malty, a color associated with a good cup of tea, but in truth our tea has quite a bland flavor compared to purer sorts. It is drunk with milk to balance the bitter flavors, and also to give us comfort during the cold and rainy days. It is a drink of basic flavors, unsophisticated and unpretentious, especially when drunk from a mug.

The porcelain cup I'm drinking tea from on my roof is the last of the set given to my mum and dad for their wedding. Since then, times have changed, and a tea set is no longer an essential part of a newly married couple's home, because sophistication and refinement are not judged by fine china and fine tea anymore. Porcelain has had to reinvent itself as modern and utilitarian. Wedding presents these days do include porcelain but usually in the form of plain white plates and even mugs, designed to look smart but most importantly to be dishwasher compatible.

I know it is my daily use of this last tea cup from the Miodownik wedding set that will ultimately spell the end for it. Every time tea is poured into it, the heat from the tea causes stresses within its frame, which pull the crack apart, while the weight of the tea within it causes a few more atomic bonds to break. Little by little, the crack increases its length, eating away at the cup from within it like a little worm. One day it will shatter into pieces. Maybe I shouldn't use it anymore but preserve it as a memento of my parents' marriage. But I prefer to think that using it every day to drink tea is a way of toasting their love for each other, which is what this cup was designed to do.

IN THE 1970S THERE was an American TV series called *The Six Million Dollar Man.* The premise of this series was that an astronaut called Steve Austin had had a terrible crash and was so close to death that it was worth trying experimental surgery to rebuild his body and sensory faculties. The procedure was designed not just to rebuild him, though. It would entirely re-engineer him, making him "Better, Stronger, Faster." The TV show didn't dwell much on the details of the complicated surgery and the implantation of bionic devices involved, but focused instead on the superhuman capabilities of the rebuilt Steve Austin, who could now run incredibly fast, jump enormous fences, and sense far-off dangers. My brothers and I loved this TV show, and we believed it. So when one day I broke my leg jumping off a climbing frame, it was with some sense of wonder and anticipation that I traveled to the hospital in our purple Peugeot 504 estate car, crammed in the back with my three brothers, all chanting in their squeaky voices: "We Can Rebuild Him, Better, Stronger, Faster . . ."

Arriving in Emergency, I was quickly and expertly examined and diagnosed. My leg was indeed deemed to be broken, but the doctors said that the natural self-healing mechanisms of my bones would repair the damage. This was disappointing news to me and sounded like a cop-out on the part of the medical establishment. Why weren't they going to rebuild me? I consulted my mother, and she confirmed that even something as hard as a bone could heal itself.

The doctors explained that bones have a soft inner core encased in a much harder outer layer, a bit like a tree; that at an invisibly small scale this inner core is made of a porous, mesh-like structure, and that this allows the cells within the bone to move about constantly, breaking down the bone and remodeling it. This is why bones, like muscles, get stronger and weaker depending on their use, building themselves up in response to the forces acting on them caused by activities such as jumping and running but mostly just by bearing a person's weight. One of the big problems for astronauts, the doctors explained to me, was the loss of bone strength that occurred when no such forces acted upon them because of the weightlessness of space. Had I been in space recently, they asked, thinking this hilariously funny. I scowled at them.

Despite the remodeling that is constantly going on in our bones, to repair a broken leg well requires that the two sides of the fracture remain in perfect contact with each other. Which, the doctors explained, meant that I would have to have a treatment to immobilize my leg for a few months – a treatment that had ancient origins, one that had been used by the ancient Egyptians and the ancient Greeks, and that was not high tech at all. It was simply to wrap my leg up in a stiff bandage.

The Egyptians used linen and the same techniques involved in mummification for this purpose; the Greeks used cloth, bark, waxes, and honey. The cast I was given, however, was made of plaster, a nineteenth-century Turkish innovation. Plaster is a ceramic made of the dehydrated mineral gypsum, which when mixed with water becomes hard like cement. Plaster is too brittle to be used on its own, though. It will simply crack into pieces after a few days. But when mixed with cotton bandages it becomes much tougher, the fibers of the bandages reinforcing the cement and stopping cracks from growing. In this form it will cocoon a broken leg for weeks. The major advantage of this over the Egyptian and Greek methods was that I didn't have to be confined to bed for the three months it would take my leg to repair itself. A plaster cast is stiff and strong enough to take the weight of a person and to withstand the knocks that occur when walking about on crutches, while allowing a perfect recovery. Until this material innovation, a broken leg often resulted in permanent lameness.

I still remember the moment that the wet plaster was applied to the bandages that had been wrapped around my leg. It was a rather weird combination of heat, caused by the reaction of the gypsum with the water, and prickliness as the soft bandages surrounding my skin grew stiff. I had a sudden itch right in the middle of my leg and had to be restrained from trying to satisfy it, which was excruciating. For the next few months, that itch would come back time and time again, usually in the middle of the night, but there was nothing I could do about it. This was the price I had to pay, my mum said, for being rebuilt like the Six Million Dollar Man. I complained that I wasn't being rebuilt—I wished I *was* being rebuilt—but instead they were merely getting my body to repair itself. I wouldn't be faster, stronger, or better than before, I'd be just the same, which wasn't very fast or strong at all. My mum told me, not unreasonably, to shut up.

My life since has been punctuated by a series of serious injuries and associated hospital visits. I haven't broken every bone in my body, but I have had a good try. I have cracked ribs and fingers, I have split my head open, I have smashed through glass, I have ripped my stomach lining, I have been stabbed; but after each incident my body healed itself, albeit under the supervision of the medical establishment. In my whole life, there have only been two problems that required doctors to "rebuild" me. The first was some time ago, but it has continued as a recurrent problem ever since.

It started with a dull ache in my mouth, which over a few days turned into a sharper, more acute pain localized in one of my teeth. Drinking hot drinks made it much worse, and then one day while biting into a sandwich I heard a horrible crunch, the sort of sound that makes your skin crawl. It was all the worse for coming from inside my mouth, and worse still that it was accompanied by an intense pain that seemed to travel like a lightning bolt through the roof of my mouth into my brain. I gingerly explored the damaged area with my tongue and to my horror discovered jagged peaks where once there had been a smooth hard tooth. It felt as if half my tooth had sheared off, which it had. After this, I couldn't eat or drink anything, because one of my nerves appeared to have been exposed by the breakage and was ultra-sensitive to anything that came into contact with it, stabbing me with pain every time this happened. My mouth now seemed to be a no-go zone, and I could think of nothing but how to stop the pain.

The Egyptians and Greeks couldn't repair this. Our ancestors lived with tooth cavities, and they lived with daily toothache. When the pain got too bad, the tooth was pulled out, either by the local blacksmith using his pliers or, if they were lucky, by a trained physician. As medicine progressed, anesthetics became available to soothe the pain, such as laudanum, a tincture of opium.

The invention in 1840 of an alloy comprised mainly of silver, tin, and mercury, called amalgam, was the turning point. In its preliminary form, amalgam is a liquid metal at room temperature because of its mercury content. However, when it's mixed with its other components, a reaction takes place between the mercury and the silver and tin that results in a new crystal, which is fully solid, hard-wearing, and tough. This miracle material could be squirted into a cavity while it was liquid, and then left to set hard. As it solidifies, the amalgam also expands slightly, wedging the filling into the cavity so that it becomes strongly mechanically bonded to the tooth. Fillings made from amalgam were far superior to equivalents made of lead and tin, which had both been tried but were too soft to last long and couldn't be poured into the cavity as a liquid without being heated to their much higher melting points, causing incredible pain in the process.

A hundred and fifty years after this alloy had first been used to treat cavities cheaply and without pulling teeth, I received my first dental amalgam filling. I still have it now, and can feel its smooth, sleek surface with my tongue. The filling transformed me from a physical and mental wreck of a boy into a bouncing, probably slightly irritating one, once more. Since then I have had eight other fillings, the first four using amalgam and the four others using a composite resin. These composite fillings are a combination of a strong transparent plastic and a silica powder that makes them hard and resistant to wear, while also matching the color of teeth better than the amalgams. These fillings, like amalgam, are molded into the cavity while liquid. Once they're in place, though, a small ultraviolet light is introduced into the mouth, which activates a chemical reaction within the resin that sets it hard almost instantly. The other modern option is the removal of the damaged tooth and its complete replacement with a porcelain (or zirconia) replica. These tend to be even harder-wearing than composite fillings and are an even better color match, too. Without these dental biomaterials, I would now have few of my own teeth left.

There is one other biomaterial that I am also reliant on to this day, which was inserted into my body in 1999 while I was working in New Mexico. My dependence came about as a result of playing soccer on an indoor court. I had the ball at my feet and was just executing a quick turn when I felt a twisting pain in my knee accompanied by a very distinct popping noise. The idea that by merely twisting my knee, without any external impact, I could mechanically tear it apart seems odd. But this is what I had done. I had snapped one of the ligaments that held my right knee in place, called the anterior cruciate ligament.

Ligaments are the elastic bands of the body. Along with our muscles and tendons, which attach our muscles to our bones, they hold our joints together and make us springy. It is the ligaments' job to connect one bone to another. They are viscoelastic, which means that they will stretch immediately a certain amount, but then if that stretch is held, they will flow and lengthen. It is part of the reason why athletes do stretching exercises to make their joints more flexible: they are lengthening their ligaments. Despite playing such a vital role in our joints, ligaments lack a blood supply, and so once they have snapped it is virtually impossible to grow them back. So, to have full use of my knee again, I would need a replacement.

There are a number of surgical approaches to this. My surgeon opted to use part of my own hamstring to remodel my anterior cruciate ligament, but in order to attach it mechanically to my knee he had to use some screws. Screws that, if I was ever to play football or go skiing again, would have to hold the replacement ligament securely in place.

Our bodies are very picky about materials inserted into them. Most things are rejected, but titanium is one of the few metals they will tolerate. More than that, titanium will undergo osseointegration, which means that it will form strong bonds with living bone. This is useful if you want to tether a piece of hamstring to a bone and be sure that the bond will not weaken and loosen with time. My titanium screws are still in place more than ten years later, and because of titanium's remarkable combination of strength and chemical inertness—there are very few metals that do not react in some way with the body; even stainless steel is not impervious to the chemical rigors of life within the body—they should be in pristine condition. Thanks to its strong surface coating of titanium oxide, titanium can last a lifetime, and I am certainly hoping it will last mine. Titanium can also withstand high temperatures, and so the screws are likely to be the last recognizable part of me once I die and am cremated. When they appear in the light again, I hope my relatives give them their due credit, as without them I wouldn't be able to do many of the things I love doing: running, playing football with my family, or walking in the mountains. The titanium screws, and my surgeon, have given me back my athleticism, for which I owe them a huge debt.

I am not dead yet, of course, and I would like to retain my physical strength and health for another fifty years. To do so I will almost certainly have to be rebuilt some more. Looking at the current technology gives me hope, because although we are a long way from having the technology of the Six Million Dollar Man, the last forty years have delivered rather impressively on this front.

On the next page is a picture of my grandfather, who died at the age of ninety-eight. He led a long life and was mentally active and able to walk, albeit with a stick, up until his death. Not everyone is so lucky. Even so, he had many health problems, and shrank in height quite considerably. Is such decline inevitable, or will we in the future be able to combat the major effects of old age by rebuilding the human body? Will the new technologies now coming

My mum walking with my grandfather in 1982.

out of biomedical research laboratories allow me to look forward to living until the age of ninety-eight while still being able to walk, run, even ski, with the same health and mobility that I have now at the age of forty-three?

In terms of mobility, the first things to wear out in the body are not the muscles or even the ligaments (I have been a little unlucky there), but the internal surfaces of the joints. Knee joints and hip joints are especially vulnerable in this respect because they are complicated moving mechanisms that bear a lot of weight, but elbow, shoulder, and finger joints also wear out. This mechanical wear and tear results in the painful and chronic condition osteoarthritis. Another type of arthritis, rheumatoid arthritis, is caused by the body's immune system attacking the joints and has a similar effect. But whether your joints destroy themselves, or whether you do it for them by crashing your car or playing sports, once you've lost the use of your hip, knee, elbow, or any other joint, no amount of rest and immobilization is going to solve the problem. Unlike the rest of your bones, the internal surfaces of your joints will not heal themselves. This is because they are not made of bone at all.

Hip joint replacements have been around for quite a while. The first attempt to replace a hip joint was in 1891 and used ivory, but titanium and ceramic are used more now. These replacement joints have been a startling success, partly because the mechanism of the hip is fairly simple: a single ball-and-socket mechanism, which allows our legs to be rotated in a whole range of ways (most of which do not come naturally—if you have ever done yoga, you will know what I mean). There is even a social ritual designed to show off hip movement, called disco dancing, and success in this area, combined with fashionable dress, can designate someone as culturally "hip."

Our hips are formed inside the womb: a ball of bone grows at the top of the femur, the bone within your thigh, and this fits perfectly inside the matching socket of your pelvis. From that point onward, these two bones grow at the same rate, ensuring that as we get bigger the joint still fits. The surfaces of these (and all) bones, though, are fairly rough, and so your body grows an outer layer of tissue, called cartilage, to line the socket at the point where the two bones touch. This tissue is softer than bone but much more rigid than muscle and creates a smooth interface between the two bones while also acting as a shock absorber. The joint is then kept together by ligaments, muscles, and tendons, which limit its movement and stop the ball being ripped out of the socket when you run, jump, and, yes, jive. When arthritis strikes, it is this cartilage that has been damaged, and this cartilage that does not grow back.

A hip replacement, then, involves sawing off the ball at the top of your femur and replacing it with a titanium ball. A new socket, tailored to fit this ball, is drilled into your pelvis and lined with a high-density polyethylene, which acts as the cartilage. These replacements restore full mobility and can last tens of years, ultimately needing replacement only when the polyethylene wears out. New versions of these artificial hip joints have been made to fit together so smoothly that the polyethylene cushion is unnecessary, but as yet it is too soon to know whether they will be longer lasting, since there may turn out to be other problems of wear caused by the direct meeting of metal on metal or, in the case of newer implants, ceramic on ceramic. Nevertheless, hip replacements are now fairly routine operations and have already helped millions of people regain their mobility in old age.

Knee joint replacements work in a similar way, except that the joint is more complicated in its mechanism: it is not a ball-and-socket joint, and yet it is required to allow both bending and twisting motions. Next time you are sitting in a café with little to do but watch the world go by, have a look at how people walk. It is kneeled—meaning that you push your knee out ahead of you, positioning it above the point where you wish to plant your next step, allowing the lower leg and foot to swing into place underneath it. Once planted, the foot has to adjust its angle to the terrain, twisting or tilting it, both of which also involve complex knee adjustments and realignment. Running is even more stressful on the knee as it must do all this while being buffeted by repeated impacts. Try walking without bending your knees and you'll see how important this joint is to mobility.

I find the prospect of wholesale replacement of my knee and hip joints in the next ten or twenty years quite an intimidating one, although if the surgery is necessary for me to be mobile, then of course I will opt for it. But ten years is a long time in medicine and materials science, and there is research going on now which may allow me to avoid it by facilitating the regrowth of my damaged cartilage within those joints.

Cartilage is a complex living material. Like a gel, it has an internal skeleton made of fibers, in its case made mostly of collagen. (Collagen is a molecular cousin of gelatin and the most common protein molecule in the human body, responsible for giving skin and other tissues their elastic firmness — which is why anti-wrinkle creams often mention the inclusion of collagen in their formulations.) Unlike a gel, though, within this skeleton there are living cells, which are responsible for creating and maintaining it. These cells are called chondroblast cells. It is now possible to grow chondroblast cells from a patient's own stem cells. However, simply injecting these into an existing joint doesn't result in the repair of the cartilage, partly because the cells cannot survive outside their homemade habitat, their collagen skeleton. In the absence of this habitat, they die. It would be like trying to start the human race again by landing Londoners on the Moon: without the infrastructure of an existing city they are mostly helpless.

What is required is the erection of a temporary structure within the joint that mimics some of the basic internal architecture of cartilage. Introducing chondroblast cells into such a scaffold, as it is called, allows them to grow and divide and increase their population, and in doing so gives them time and space to rebuild their habitat, and so regrow cartilage. The neat thing about this scaffolding approach is that either the cells themselves can consume the scaffold or it can be designed to dissolve once the cells have finished building their habitat, leaving pristine cartilage within the knee or hip.

The idea of rebuilding cartilage tissue using such a scaffold may seem far-fetched but it is an established method, pioneered in the 1960s by Professor Larry Hench. He was challenged by an army colonel to find a way to help regenerate the bones of Vietnam War veterans which would otherwise be amputated: "We can save lives but we cannot save limbs. We need new materials that will not be rejected by the body." Hench and others searched for materials that could be a better match for bone, and discovered a material called hydroxylapatite, a mineral which occurs in the body and bonds very strongly to bone. They experimented with many formulations and in the end found that when it was made in the form of a glass it had extraordinary properties. This bioactive glass was found to be porous, meaning it contains tiny channels. Bone cells, called osteoblasts, liked to inhabit these channels, and as they created new bone they would break down the bioglass around them, as if they were eating it.

Such tissue engineering has been very successful and is now used to provide synthetic bone grafts, and to rebuild the bones of the skull and face. It is not yet used on more structural bones, which must hold weight, since it takes considerable time to rebuild the bones and the scaffolding alone can't withstand large stresses while this is taking place. The current strategy for the building of these larger structures is to do so in the laboratory, since the scaf-

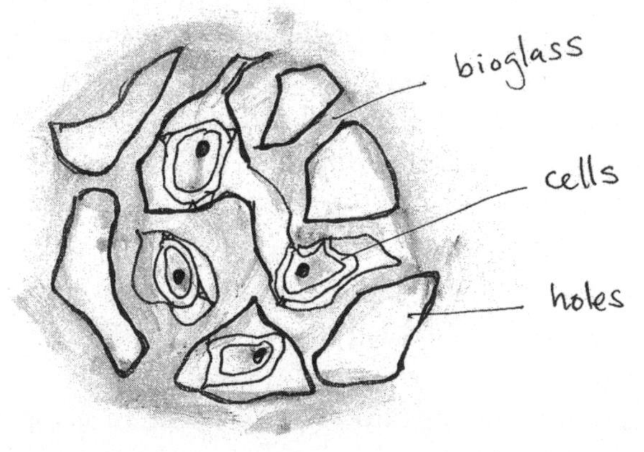

Bioglass scaffold material with cells growing inside the structure.

folding process works not just inside the body but outside it too. In this case, the cells must be nurtured in a bioreactor, which mimics the temperature and humidity of the human body while also providing the cells with nutrients. The success of this technology has opened up the possibility of building fully functioning replacement body parts in their entirety. The first steps in this direction have already been taken, with the successful development of a human windpipe grown in a laboratory.

The project started with a patient with a diseased windpipe, which needed to be removed because it was cancerous. Without a replacement, the patient would require mechanical aids to breathe for the rest of his life. The first step was to scan the patient using an x-ray technique common in hospitals called a CAT scan. CAT scans are often used to find cancerous lumps in the brain and other organs. But in this case the CAT scan was used to provide a 3D image of the patient's windpipe. This image was then taken to a 3D printer, a new type of manufacturing technique that creates whole objects from digital information. The way it works is not so unlike a normal printer, except that instead of expressing dots of ink on to a page, the print head releases blobs of material, depositing one layer of the object at a time and gradually building the object layer upon layer. The technique can now be used to print not just simple objects, such as cups and bottles, but more complex objects with moving parts, such as hinges and motors. At present the technology can print in a hundred different materials, including metals, glass, and plastic. Using a 3D printer, Professor Alex Seifalian and his team created an exact replica of his patient's windpipe made from a special scaffold material they had developed, one that was tailor-made to accommodate the patient's stem cells.

The role of adult stem cells is to renew our tissues, and each type of cell has an equivalent stem cell to produce it. The stem cells that produce bone cells are called mesenchymal stem cells. Having built the scaffold, Professor Seifalian's team implanted it with mesenchymal stem cells taken from the patient's bone marrow and placed the whole object into the bioreactor. These stem cells then turned into a range of different cells that started to build cartilage and other structures, creating a living, self-sustaining cellular environment, while at the same time dissolving the scaffold around them. Eventually, all that was left was a new windpipe.

One of the major advantages of this technique is that the implant is made of the patient's own cells, and once implanted becomes part of his body. The patient therefore does not need to take any immunosuppressant drugs, which have significant side effects, to stop the body from rejecting the transplant. (Suppressing the immune system in order to protect the implant can make the patient much more vulnerable to infections of all kinds, as well as to parasites.) However, for the treatment to be effective, the body has to develop a blood supply to the windpipe, and for the moment it remains to be seen quite how well the body develops these supply connections. The cell ecology of the windpipe must also remain stable if the windpipe is to retain its shape and allow the patient to breathe normally. A further problem is that of sterilization. The polymers from which the scaffold is printed are delicate, and they cannot survive the high temperatures of traditional sterilization. Nevertheless, despite all these challenges, the first windpipe transplant made from a patient's own stem cells was completed on July 7, 2011.

The windpipe scaffold, developed by Professor Seifalian's team, with stem cells incorporated before transplantation.

The success of this technology has accelerated progress toward the production of a new generation of scaffold materials. A windpipe has to be mechanically functional and to develop a blood supply to survive long term, but it isn't an organ with a regulatory role in the body. The next challenge is to grow livers, kidneys, and even hearts. At the moment, if you lose the function of any of these major organs, then a transplant is needed to restore you to full health. Such transplants require donor organs to be healthy and biologically matched to you, and also that you remain on drugs for the rest of your life to prevent the rejection of the organ. But because, in most cases, donor organs are the only hope for patients ever to regain the health and independence they once enjoyed, they are in short supply.

This chronic shortage is having three effects. First, it means that patients without liver or kidney function need long-term health care, which is very expensive and robs them of their independence. Second, patients waiting for heart transplants often die before a suitable organ becomes available. Finally, there is a growing black market for organs, which means that poor people, especially in developing countries, are coming under increasing pressure to sell their organs. This practice has been documented in several studies, most recently by one at Michigan State University, which showed that thirty-three kidney sellers in Bangladesh didn't get the money they were promised and suffered severe health problems as a result of their surgery. Typically, the practice involves being flown to another country and taken to a private hospital where the wealthy recipient awaits the organ and where the procedure is carried out. The average price of a kidney is quoted at $1,200.

These problems are not going to go away unless an alternative treatment to organ transplants is developed. Tissue engineering using biomaterial scaffolds is currently the most promising alternative technology. Clearly the challenges are enormous. These organs have complex internal structures and often contain many different types of cells, which all interact to perform the organ function. In the case of livers and kidneys, they need not only to develop a blood supply but also to be connected to the main arteries of the body. The heart is a particularly acute problem because we have only one, and without one that functions, we die. Types of artificial heart have been developed, but the longest anyone has survived with one is a year.

It is likely that 3D printing will play a central role in any technology that involves engineering new organs. These 3D printers are already used widely for making dental implants, and in 2012 this same technology was used to create an artificial jawbone for an eighty-three-year-old woman. This jaw was made from titanium, but the printing of a scaffold material to accommodate the cells that will turn it into a patient's own bone is rapidly becoming possible.

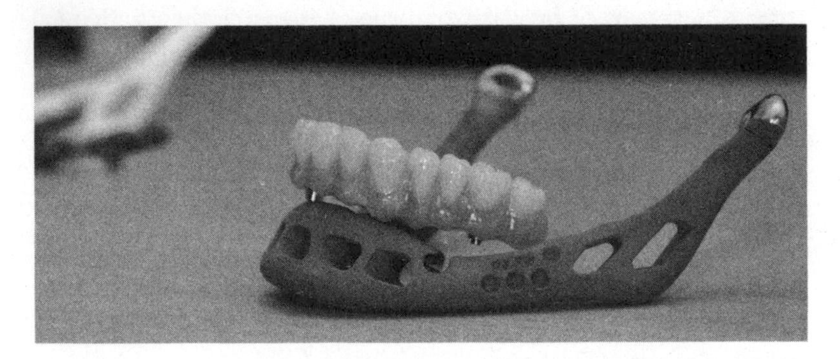

Artificial jawbone created using a 3D printer.

All the pieces of the puzzle seem to be in place for a wholesale re-engineering of the major organs of the human body, and so it doesn't seem too far-fetched to think that by the age of ninety-eight I may have a new heart, a few other replacement organs, and new joints keeping me fit and healthy. But will I be like the Six Million Dollar Man, "Better, Stronger, Faster"?

It's hard to say, but the answer is probably no. This is because much of what makes us older is not the age of our cells but the deterioration of the systems that generate them. Aging is the cellular equivalent of playing telephone: each generation of cells does not quite regenerate the structure it inherited and so mistakes and imperfections creep in. My skin has aged not because my skin cells are forty-three years old—they are not; they are constantly being replaced with new cells generated by my adult stem cells—but because, over time, problems and imperfections have developed in the structure of my skin and have then been passed from one generation of cells to the next. Spots form, the skin thins, wrinkles appear. These problems will continue to be reproduced.

The same is true of the cardiovascular system. Circulatory diseases account for almost a third of all deaths in the UK, more than any other cause of death. In other words, I am most likely to die from a heart attack or a stroke. This is essentially a mechanical failure of the cardiovascular system, the network of heart, lungs, arteries, and veins that keeps my body going. But while surgeons have become very good at patching the system, replumbing it when it goes wrong, and even replacing parts of it with transplants (or artificially grown implants), this doesn't change the fact that the whole system has seen a lot of use. A ninety-eight-year-old cardiovascular system which has been repaired is still ninety-eight years old and will become more and more vulnerable to failure. For the foreseeable future, replacing the whole vascular system is not in the cards at all.

The upshot of this is that although it will become increasingly effective to grow and replace body parts, the interconnections between these different organs and the thousands of different systems on which our bodies depend will continue to accrue defects that impair performance. We will still get old.

Synthetic implants are a radical solution to the problem of some parts of the body wearing out before the rest, but they are not a solution to the ultimate problem (if we can call it that) of death: what they offer is a better kind of life. Already robotic limbs have been developed to replace those lost by amputees. These electro-mechanical devices pick up the nerve impulses delivered to the missing body part by the brain and translate them into the equivalent hand grips or leg movements in the artificial limb. The same technology has now been used to help those who are paralyzed from the neck down to control robotic limbs and regain a measure of independence. Although these technologies are designed for those disabled or paralyzed, they could also be used by someone who has lost movement as a result of aging.

This is a different kind of future from that offered by tissue engineering: it is a bionic future where our mobility and material connection to the world become more and more mediated by synthetic and electronic components. This is the technology that was envisaged in *The Six Million Dollar Man,* which allowed him to be "Better, Stronger, Faster." In today's money, those six million dollars would be thirty-five million dollars, and although this is a fictional figure, it does highlight an important truth about life-extending technologies: they are expensive. The technology that will allow us to lead fit and healthy lives up to the age of a hundred is likely to cost a great deal of money. Who will pay? Will it be a luxury? Will it be only the rich who can play tennis at the age of ninety-eight while the rest of us are in wheelchairs? Or will the technology simply allow us to work longer, making it normal to do so until eighty or ninety? I prefer the latter future, but if thirty-five million dollars is anywhere close to the right figure, then most of us will never be able to afford it, however many years we work.

I will in all likelihood live to be ninety-eight years of age. Whether I will have shrunk to half my height and be walking slowly with a stick like my grandfather before me, or whether I will be playing tennis and football with my grandchildren, will be as much due to brilliant biomaterials research as to how the economics of medicine pan out. I do hope, though, that the chant of my brothers all those years ago—"We Can Rebuild Him, Better, Stronger, Faster"—does come true. I could handle a little bit of immortality.

11 SYNTHESIS

IN THIS BOOK I have delved into our material world in an attempt to show that although the materials around us might seem like blobs of differently colored matter, they are in fact much more than that: they are complex expressions of human needs and desires. And in order to create these materials—in order to satisfy our need for things like shelter and clothes, our desire for chocolate and the cinema—we have had to do something quite remarkable: we have had to master the complexity of their inner structure. This way of understanding the world is called materials science, and it is thousands of years old. It is no less significant, no less human, than music, art, film, or literature, or the other sciences, but it is less well known. In this final chapter I want to explore the language of materials science more fully, because it offers a unifying concept that encompasses all materials, not just the ones we have considered in detail in this book.

This unifying concept is that although a material may look and feel monolithic, although it may appear to be uniform throughout, this is an illusion: materials are, in fact, composed of many different entities that combine to form the whole, and these different entities reveal themselves at different scales. Structurally, any material is like a Russian doll: it is made up of many nested structures, almost all of which are invisible to our eyes, each one smaller and fitting exactly into the one before. It is this hierarchical architecture that gives materials their complex identities—and, in a very literal sense, it also gives us our identities too.

One of the most fundamental of these material structures is the atom, but it is not the only structure of importance. At the larger scales there are dislocations, crystals, fibers, scaffolds, gels, and foams, to name a few that have been featured in this book. Taken in isolation, these structures are like characters in a story, each contributing something to its overall shape. Sometimes one character dominates the story, but it is only when they are put back together that they explain fully why materials behave the way they do. As we have seen, the reason why a stainless steel spoon doesn't taste of anything is because the chromium atoms within its crystals react with oxygen in the air to form an invisible protective layer of chromium oxide on the surface. If you scratch its surface, this protective layer grows back more quickly than rust will form. This

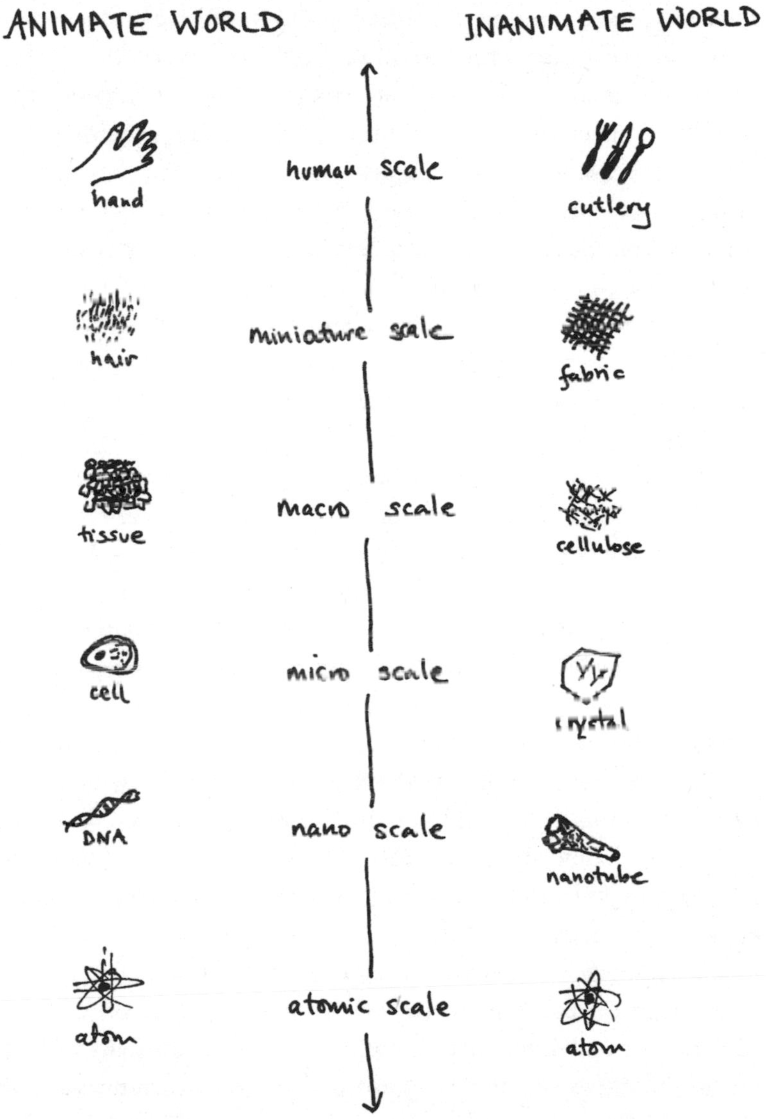

ANIMATE WORLD

INANIMATE WORLD

hand

human scale

cutlery

hair

miniature scale

fabric

tissue

macro scale

cellulose

cell

micro scale

crystal

DNA

nano scale

nanotube

atom

atomic scale

atom

is why we are the first generations not to taste our cutlery. Such molecular explanations are satisfying, but in this case they only account for one aspect of the material: its tastelessness. A full understanding of why stainless steel behaves the way it does requires you to consider all of the structures of which it is composed.

When you start to look at materials in this way, you soon realize that all materials have a common set of structures within them. (To take the simplest of examples, all materials are made of atoms.) And before long, you'll find that metals have much in common with plastics, which in turn have much in common with skin, chocolate, and other materials. In order to visualize this connection between all materials, we need a map of this Russian doll–like material architecture: not a normal map that shows the variety of terrain on a single scale, but a map that shows terrain on a variety of scales: the inner space of materials.

Let's start with the primary ingredients: the atoms. These are approximately ten billion times smaller than us, and so structures at the atomic scale are obviously invisible to our eyes. On Earth, ninety-four different types of atoms naturally exist, but eight of these elements make up 98.8 percent of the mass of the Earth: iron, oxygen, silicon, magnesium, sulfur, nickel, calcium, and aluminum. The rest are technically trace elements, including carbon. We have the technology to transform some of the common ones into the rare ones, but this requires a nuclear reactor, which costs even more money than mining and results in radioactive waste. This is essentially why gold is still valuable in the twenty-first century. If gathered together, all the gold ever mined would fit inside a large town house. Nevertheless, the rarity of certain types of atoms on the planet, such as neodymium or platinum, which are so technologically useful, may not ultimately be a problem, because a material is not defined by its atomic ingredients alone. As we now know, the difference between hard transparent diamond and soft black graphite is not to do with their atoms: in both cases, they are made of exactly the same pure element, carbon. It is by changing how they are arranged, by altering them from a cubic structure into layers of hexagonal sheets, that the radical differences in their material properties are brought about. These structures are not arbitrary—you cannot create any structure—but are governed by the rules of quantum mechanics, which treat atoms not as singular particles but as an expression of many waves of probability. (This is why it makes sense to refer to the atoms themselves as structures, as well as their formation when they bond with one another.) Some of these quantum structures create electrons that can move, and this results in a material that can conduct electricity. Graphite has such a structure, and so conducts electricity. Exactly the same atoms in a diamond but in a different structure do not allow the electrons to move so easily within the crystal, and so diamonds do not conduct electricity. It is also why they are transparent.

This apparent alchemy illustrates that even with a very restricted set of atomic ingredients you can create materials with wildly different material properties. Our bodies are very good examples of this: we are mostly made of carbon, hydrogen, oxygen, and nitrogen, and yet through subtle rearrangements of the molecular structure of these ingredients, and the sprinkling of a few minerals such as calcium and potassium, an immense diversity of biomaterials results, from hair, to bone, to skin. It is hard to overestimate the philosophical as well as the technological importance of this dictum of materials science: that knowing the basic chemical composition is not enough to understand materiality. It is, after all, what makes the modern world possible.

To make any material, then, we need to bond atoms together. If you assemble a hundred or so of them, you have what is called a nanostructure. "Nano" means "a billionth," and this world of the nanoscale features things that are roughly a billion times smaller than us. This is the scale of macromolecules, where tens and hundreds of atoms get together to form much larger structures. These include the proteins and fats in our bodies. They also include the molecules at the heart of plastics, such as the cellulose nitrate used to make celluloid, or the lignin that is removed from wood to make paper. Holes in the structure at this scale create a fine foam, such as that of aerogel. These are all structures that feature in this book under different guises. What unites them is that they express their characteristics at the nanoscale, and it is manipulations at this scale that will affect their properties. Humans have been controlling the nanoscale for thousands of years but only indirectly, using chemistry or using metallurgy in a hearth. When a blacksmith hits a piece of metal he or she is changing the shape of the metal crystals within it by "nucleating" nanoscale dislocations—in other words, by causing the transfer of atoms from one side of the crystal to the other at the speed of sound. We don't see these nanoscale mechanisms, of course. At our scale we simply see the metal changing shape. Which is why we perceive the metal to be monolithic and slab-like: all the intricate mechanics of crystals have been beyond our comprehension until very recently.

The reason why *nanotechnology* is such a buzzword today is that we now have microscopes and tools for directly manipulating structures at this scale and so creating a vastly larger array of such nanostructures. It is now possible to create structures at this scale that will collect light and store it as electricity, to create light sources, and even to create nanoparticles that can sense smells. The possibilities seem limitless, but what is more interesting is that many of the structures at this scale self-assemble. That means that the materials are able to organize themselves. This might seem spooky but is perfectly in line with the existing laws of physics. The crucial difference between the car motor and a nano-motor is that in the case of the nano-version the physical forces that dominate at that scale, such as electrostatic and surface tension forces, which can pull things together, are very strong, while gravitational forces are very weak. At the scale of a car, by far the strongest force is the gravitational force of the Earth, which pulls the various bits of the motor apart. The result is that nano-machines can be designed to assemble themselves using electrostatic and surface tension forces (and heal themselves in the same way). Much of this molecular machinery already exists inside cells, which is how they can assemble themselves, whereas at the human scale we need things like muscles and glue.

Nanostructures are still far too small to see or even feel, so to integrate them into a material object that we can interact with, we have to group them together and connect them into microscopic structures, which are ten to a hundred times bigger, but nevertheless still invisible. This is the scale at which we encounter one of the greatest technological triumphs of the twentieth century: the silicon chip. These chips are tiny collections of silicon crystals and electronic conductors, and they are the basic engine of the electronic world. There are billions of them inside the many electronic machines that surround us—they play our music, they take our holiday photos, and they wash our clothes. They are the man-made equivalent of the neurons in our brains and exist at the same scale as the nucleus within our cells. Strangely enough they contain no moving parts, using only the electric and magnetic properties of materials to control the flow of information.

This is also the scale of biological cells, of iron crystals, of the cellulose fibers of paper and the fibrils of concrete. At the same scale still we find another remarkable man-made structure: the chocolate microstructure. Here the six types of cocoa butter crystal structure, each with a different melting temperature, create very different textures in the chocolate. At this scale too are the crystals of sugar and the grains of cocoa solids containing the flavor molecules of chocolate. Controlling this microstructure controls the taste and texture of chocolate, and this is a large part of the craft of the chocolatier.

At this microscale, materials scientists are starting to design structures that are able to control light. These so-called metamaterials can be formed with variable refraction indices, which means they can bend light any way they want to. This has yielded the first generation of invisibility shields, which when surrounding an object bend light around it so that from whichever direction you try to observe it, it appears to vanish.

The macroscale binds together the atomic structures, the nanostructures, and the microstructures. It is just on the edge of what we can see. The touch screen of a smartphone is a good example of such a structure. It looks smooth and without detail, but if you put a drop of water on the screen it acts as a lens and allows you to see that it is in fact made of tiny individual pixels of red, green and blue. All of these tiny liquid crystals can be individually controlled, allowing them to combine at the human scale to represent all of the colors in the visual spectrum, and they can be switched on and off fast enough to make it possible to watch movies. Porcelain is another good example of the effect of changes at the macroscale: it is here that all the different glass and crystal structures combine to produce a strong, smooth, and optically dynamic material.

The miniature scale combines the atomic structures, nanostructures, microstructures, and macrostructures into a structure that is just visible to the naked eye. This is the scale of a piece of thread or a strand of hair, the scale of a needle and the line width of this font size. When you look at and feel the grain of wood, you are seeing and feeling the combination of all these structures at the miniature scale. This combination gives wood its characteristic feel of being stiff but not too hard, of being light and warm. Similarly, ropes, blankets, carpets, and most importantly clothes are made at this scale, and the strength, flexibility, smell, and feel of these materials are the result at this scale of the combination of all structures embodied within them: a thread of cotton may look superficially similar to a thread of silk or Kevlar, but it is the hidden detail of their atomic-, nano-, micro-, macro-, and miniature structures that makes the difference between something that can protect against a knife or feel as smooth as cream. It is at this scale that our sense of touch engages with materials.

Finally we reach the human scale, the scale at which all of the preceding structures combine and at which we encounter the stuff we can hold in our hands, inhabit with our bodies, or put by the forkful into our mouths. This is the scale of sculpture and art, of plumbing and cooking, of jewelry and building construction. Materials at this scale are recognizable objects, such as plastic pipes, tubes of oil paint, lumps of stone, loaves of bread, and metal bolts. At this scale they look once more like monolithic lumps of matter, but we have seen that this is not the truth. But because it is only with magnification that the hidden depths of these lumps of matter are revealed, it has taken us until the twentieth century to discover this multi-scale architecture that inhabits all of the stuff around us. It is this that explains why all metals may look the same but behave very differently, why some plastics are stretchy and soft while others are hard, and how we can turn sand into skyscrapers. It is one of the proudest achievements of materials science, because it explains so much.

Although designing structures at different scales has allowed us to design new materials, the real challenge of the twenty-first century is to link up designed structures at all of these scales into a macroscopic human-sized object. Although smartphones are an example of such integration, combining a macroscale touch-sensitive screen with nanoscale electronics, the possibility that whole objects might be wired up throughout, as if permeated by an entire nervous system, is now becoming conceivable. And if we can achieve this, then one day whole rooms, buildings, perhaps even bridges may generate their own energy, funnel it to where it is needed, detect damage, and self-heal. If this seems like science fiction, bear in mind that it is only what living materials do already.

Since all of the small scales of a material are encapsulated within the larger ones, as things increase in size they become more complex. This means that the world of subatomic particles and quantum mechanics, although often perceived to be the most complex part of science, is in reality much less complex than, say, a petunia. This has long been recognized by biologists and physicians, whose science has been driven by empirical and experimental methods (rather than theoretical ones) for so long precisely because the organisms they study, being large and living, are so complex as to defy theoretical description. But as the scale chart shows, living matter is, in some sense, no different conceptually from non-living matter. What dramatically distinguishes the two is that in living materials we find there is an extra degree of connectivity between the different scales: living materials actively organize their internal architecture. They do this by setting up communication between the different scales of the organism. In a non-living material, a mechanical stress imposed at the human scale has all sorts of effects at different scales, causing many internal mechanisms to react in response: as a result it might change shape or break or resonate or stiffen. A living material, on the other hand, can detect that such a stress is occurring and adopt a course of action in response: it might push back, or it might instruct the whole organism to run away. Obviously there is a vast range of such animated behaviors: the branch of a tree behaves passively, as if it were an inanimate material, most of the time, while the leg of a cat is most definitely animate most of the time. One of the biggest questions in science is whether communication between the scales combined with active responses is a sufficient explanation of what makes something alive. Such a hypothesis is not meant to downgrade the importance of living beings but rather to upgrade inanimate materials: they are much more complex than they appear.

However rapid the pace of change in our technology has been until now, the fundamental arrangement of materials on the planet has not altered. There are living things which we call life, and there are non-living things which we call rocks, tools, buildings, and so on. As a result of our greater understanding of matter, though, this distinction is likely to become blurred as we usher in a new era of materials. Bionic people with synthetic organs, bones, and even brains will be the norm.

What makes us human, though, is not just the physical materiality of our bodies, synthetic or not. We inhabit an immaterial world, too: the world of our minds, our emotions, and our perceptions. But the material world, although separate, is not entirely divorced from these worlds — it strongly influences them, as anyone knows: sitting on a comfy sofa affects our emotional state in a very different way from sitting on a wooden chair. This is because for humans, materials are not just functional. The early archaeological evidence shows that as soon as we developed tools we also started creating decorative jewelry, pigments, art, and clothing. These materials were developed for aesthetic and cultural reasons, and this has been a strong driver of materials technology throughout history. Because of this strong connection between the materials and their social role, the materials that we favor, the materials that we surround ourselves with, are significant to us. They mean something, they embody our ideals, they give us part of our identity.

These material meanings are embedded in the fabric of our world and overlap with their utility. Metal is very tough and strong, so it makes sense to build machines with it, but the reliability and resilience associated with metals is also used consciously by designers to give these qualities to their products. The look of metal is part of the language of industrial design: it speaks of the Industrial Revolution that first gave us mass transport and the machine age. The fact that we can mass-produce metals and shape them is part of who we are. We admire this material because it is our reliable, tough, and mechanically strong workhorse; we all rely on it every time we get in a car or train, every time we put our clothes in the washing machine, every time we shave our bodies.

Because we have a long history, our material culture is complicated. For the very same reasons that we admire metal — its associations with industry, for example — we may also dislike it. Materials have multiple meanings, and so my choice of the various adjectives that head the chapters is not meant to be definitive. These are personal choices, and each chapter is written from a personal perspective, in order to illustrate that we all have personal relationships with our material world, and these are simply mine.

We are all sensitive to the meanings of materials, whether consciously or subconsciously. And since everything is made from something, these meanings pervade our minds. We are being bombarded with them constantly by our environment. Whether we are on a farm or in a city, on a train or a plane, in a library or a shopping mall, they affect us. Of course, designers and architects consciously use these meanings to create clothes, products, and buildings that we like, that we identify with, that we want to surround ourselves with. In this way the meanings of materials are reinforced by our collective behavior and so take on a collective meaning. People buy clothes that reflect the type of person they want to be, or aspire to be, or are forced to be—fashion designers are expert in these meanings. But in every aspect of our lives we choose materials that reflect our values, in our bathrooms, in our living rooms, in our bedrooms. Similarly, others impose their values on us in the workplace, in our cities, and in our airports. There is a continual reflection, absorption, and expression going on in the material world that constantly remaps the meanings of materials around us.

This mapping, though, is not a one-way street. The desire for, say, stronger, more comfortable, waterproof, breathable fabrics creates a need for the understanding of the internal material architectures that are required to create them. This drives our scientific understanding, and so drives materials science. In a very real way, then, materials are a reflection of who we are, a multi-scale expression of our human needs and desires.

A final look at the picture of me on my roof. I hope, as a result of reading this book, you'll see it a little differently . . .

ACKNOWLEDGMENTS

From an early age my curiosity was nurtured by having a dad who was a scientist: a man who brought home bottles of acid marked DANGER; who performed experiments in his workshop in the cellar; and who bought one of the first Texas Instruments calculators. I have three brothers: Sean, Aron, and Dan. As children we explored the world in a very tactile manner, building, digging, breaking, poking, and leaping. This was all done under the beneficent eye of my mum, who approved of fresh air, eating, and combed hair. All the boys, including me, went prematurely bald as young men, so we could not please her in later life with our neat hairstyles; but we do all love cooking and this is a tribute to her. I feel great sadness that she died in December 2012 and that she never saw this book in print.

My materials science education began in earnest in the Materials Department at Oxford University, and I want to thank all the faculty and staff, in particular my tutors John Martin, Chris Grosvenor, Alfred Cerezo, Brian Derby, George Smith, Adrian Sutton, Angus Wilkinson, and, of course, the head of department, Peter Hirsch. I learned a great deal from Andy Godfrey, with whom I shared an office as a PhD student.

When I left Oxford in 1996 I moved first to the USA to work in Sandia National Laboratories, then to University College Dublin to work in the Mechanical Engineering Department, then to King's College London, and finally to University College London, where I now work. There are many people from whom I learned important stuff along the way. I am particularly indebted to Elizabeth Holm, Richard LeSar, Tony Rollett, David Srolovitz, Val Randle, Mike Ashby, Alan Carr, David Browne, Peter Goodhew, Mike Clode, Samjid Mannan, Patrick Mesquida, Chris Lorenz, Vito Conte, Jose Munoz, Mark Lythgoe, Aosaf Afzal, Sian Ede, Richard Wentworth, Andrea Sella, Harry Witchel, Beau Lotto, Quentin Cooper, Vivienne Parry, Rick Hall, Alom Shaha, Gail Cardew, Olympia Brown, Andy Marmery, Helen Maynard-Casely, Dan Kendall, Anna Evans Freke, David Dugan, Alice Jones, Helen Thomas, Chris Salt, Nathan Budd, David Briggs, Ishbel Hall, Sarah Conner, Kim Shillinglaw, Andrew Cohen, Michelle Martin, Brian King, Deborah Cohen, Sharon Bishop, Kevin Drake, and Anthony Finklestein.

My appreciation of the subject has been expanded by working with some great organizations to put on events and exhibitions, and to make programs about materials. I would like to thank the Cheltenham Science Festival, the Wellcome Collection, Tate Modern, the V&A, the Southbank Centre, the Royal Institution, the Royal Academy of Engineering, the BBC Radio 4 Science Unit, and the BBC TV Science Department.

The UCL Institute of Making is a very special place, an intellectual home, and I want to thank the whole team for their friendship and support while I wrote this book: Martin Conreen, Elizabeth Corbin, Ellie Doney, Richard Gamester, Phil Howes, Zoe Laughlin, Sarah Wilkes, and Supinya Wongsriruksa.

I want to thank those who have seen and commented on particular chapters. They are Phil Purnell, Andrea Sella, and Steve Price.

There are those who not only commented on the book as it took shape, but also encouraged me along the way. My huge thanks to my great friend Buzz Baum; my dear dad, brothers, sisters-in-law, niece, and nephews; and the members of Enrico Coen's Perugia 2012 research workshop.

This book would not have happened at all if it were not for my literary agent, Peter Tallack, my British editor, Will Hammond, and the vision and commitment of my US editor, Courtney Young, and the Houghton Mifflin Harcourt team.

Finally, this book was written in the period between my son, Lazlo, and my daughter, Ida, being born. They and their mother, Ruby, are the creative power that flows between the pages.

PHOTO CREDITS

Page 4: Central European News.

Page 37: Alistair Richardson.

Page 39: OK! Syndication/www.expresspictures.com.

Page 46: Roger Butterfield.

Page 57: Network Rail.

Page 64: Foster and Partners.

Page 70: Courtesy of Italcementi Group.

Page 88: Cadbury's.

Page 97: NASA.

Page 99: NASA.

Page 107: NASA.

Page 143: A. Carion.

Page 144: John Bodsworth.

Page 208: University College London.

Page 210: University of Hasselt.

FURTHER READING

Philip Ball, *Bright Earth: The Invention of Colour*, Vintage (2008).

Rodney Cotterill, *The Material World*, CUP (2008).

Michael Faraday, *The Chemical History of a Candle*, OUP Oxford (2011).

Stephen Fenichell, *Plastic: The Making of a Synthetic Century*, HarperCollins (1996).

Adrian Forty, *Concrete Culture: A Material History*, Reaktion Books (2012).

J. E. Gordon, *New Science of Strong Materials: Or Why You Don't Fall Through the Floor*, Penguin (1991).

——, *Structures: Or Why Things Don't Fall Down*, Penguin (1978).

Philip Howes and Zoe Laughlin, *Material Matters: New Materials in Design*, Black Dog Publishing (2012).

Chris Lefteri, *Materials for Inspirational Design*, Rotovision (2006).

Primo Levi, *The Periodic Table*, Penguin, new edition (2000).

Gerry Martin and Alan Macfarlane, *The Glass Bathyscape: How Glass Changed the World*, Profile Books (2002).

Harold McGee, *McGee on Food and Cooking: An Encyclopedia of Kitchen Science, History and Culture*, Hodder & Stoughton (2004).

Matilda McQuaid, *Extreme Textiles: Designing for High Performance*, Princeton Architectural Press (2005).

Cyril Stanley Smith, *A Search for Structure: Selected Essays on Science, Art and History*, MIT Press (1981).

Arthur Street and William Alexander, *Metals in the Service of Man*, Penguin (1999).

INDEX

Note: Illustrations are indicated by *italics*